THE STATISTICAL PHYSICS OF DATA ASSIMILATION AND MACHINE LEARNING

HENRY D. I. ABARBANEL
University of California, San Diego

**CAMBRIDGE**
UNIVERSITY PRESS

CAMBRIDGE
UNIVERSITY PRESS

University Printing House, Cambridge CB2 8BS, United Kingdom

One Liberty Plaza, 20th Floor, New York, NY 10006, USA

477 Williamstown Road, Port Melbourne, VIC 3207, Australia

314–321, 3rd Floor, Plot 3, Splendor Forum, Jasola District Centre, New Delhi – 110025, India

103 Penang Road, #05–06/07, Visioncrest Commercial, Singapore 238467

Cambridge University Press is part of the University of Cambridge.

It furthers the University's mission by disseminating knowledge in the pursuit of
education, learning, and research at the highest international levels of excellence.

www.cambridge.org
Information on this title: www.cambridge.org/9781316519639
DOI: 10.1017/9781009024846

First published 2022

Printed in the United Kingdom by TJ Books Limited, Padstow Cornwall

A catalogue record for this publication is available from the British Library.

Library of Congress Cataloging-in-Publication Data
Names: Abarbanel, H. D. I., author.
Title: The statistical physics of data assimilation and machine learning /
Henry D.I. Abarbanel.
Description: Cambridge ; New York, NY : Cambridge University Press, 2022. |
Includes bibliographical references and index.
Identifiers: LCCN 2021025371 (print) | LCCN 2021025372 (ebook) | ISBN
9781316519639 (hardback) | ISBN 9781009024846 (ebook)
Subjects: LCSH: Statistical physics–Data processing. | Stochastic
processes. | Supervised learning (Machine learning)–Mathematical
models. | Discrete-time systems.
Classification: LCC QC174.8 .A225 2022 (print) | LCC QC174.8 (ebook) |
DDC 530.13–dc23
LC record available at https://lccn.loc.gov/2021025371
LC ebook record available at https://lccn.loc.gov/2021025372

ISBN 978-1-316-51963-9 Hardback

Contents

Preface

This book explores methods for performing the tasks in Data Assimilation, a critical practical step in the transfer of information from observed data collected during measurements of a physical or biological dynamical system to a nonlinear dynamical model proposed for the that system.

The name Data Assimilation emerged over the years in the context of Numerical Weather Prediction in meteorology, but has found the same challenge in many fields of study: including in numerical weather prediction and quantitative aspects of neurobiology, as examples, and many other areas of science and technology where one must address this transfer of information as well.

Further, as we proceed through this book we show that the same questions and associated tools for answering them appear in the **equivalent** problem of Supervised Machine Learning.

These seemingly quite different formulations of questions and tools for addressing them all appear in the framework of Statistical Physics. If you have not had experience with Statistical Physics, this volume will introduce you to much of its strength without any undue suffering on your part–that's the plan anyway!

In our discussions these problems are placed into a common path integral formulation (Zinn-Justin (2002); Hochberg et al. (1999); Abarbanel (2013)) which provides a unification of critical questions and a framework in which to view methods developed in various disparate fields as they apply to many others.

What does it mean to transfer information in data to a model of the dynamical system generating those data?

In these problem areas we have dynamical equations (the model) having state variables {voltages, $V(t)$, density or concentrations of chemical constituents, $\rho(x, y, x, t)$, velocities, $\mathbf{v}(x, y, z, t)$, and so forth} as well as time independent parameters such as {viscosities, conductivities, and so forth}, any of which may be *unobserved or unobservable* in the collection of the data.

Figure 0.1 Monsieur Proust's Collection of Books: "Remembrance of Things Past"

Over some observation window in time $[t_0, t_{final}]$ we collect information on the observable variables, then use data assimilation tools to estimate the unobservable state variables and time independent parameters. With estimates of the full set of state variables at t_{final} and all the parameters, we can use these as initial conditions at t_{final} to predict forward $t > t_{final}$ as a test or validation or generalization of the model.

This volume is intended for data scientists, physical scientists, and life scientists who wish to utilize machine learning methods (Goodfellow et al. (2016); Abarbanel et al. (2018)), to simplify or accelerate calculations within their inquiries (Pathak et al. (2018); Ott (2019)).

It is intended for scientists and engineers with experience in methods of statistical physics, typically covered in beginning graduate courses in Physics and Chemistry, for transferring information in observed data to models of the observed processes with the goal of estimating unknown fixed parameters in the models as well as *unobserved state variables* in the models.

This transfer process is called by various names in different fields of study. We adopt the designation **data assimilation** following the terminology used in numerical weather prediction (Ghil and Malanotte-Rizzoli (1991); Pires et al. (1996); Kalnay (2003); Lorenc and Payne (2007); Evensen (2009); Reich and Cotter (2015)) over many decades. It is also widely referred to as state and parameter estimation in many engineering applications. These are just a few: Dochain (2003); Horváth and Manini (2008); Lei et al. (2017).

Collecting the data takes quite skilled personnel.

- The data is always noisy and typically sparse in the sense that only a (usually quite small) subset of the dynamical variables in the system producing the data are observable.
- Formulating a model for the dynamics of that system is not algorithmic. It takes experience and insight into the physical or biological or other mechanisms identified to be operating within the observed system.
- Transferring the information residing in the data to critical aspects of the selected model also takes skill.

This book is primarily about the last of these items, especially when the data are noisy and the model has errors.

An overview of these steps is this:

> In some time interval $[t_0, t_{final}]$ (or many such time intervals) we measure L quantities $\mathbf{y}(\tau_k) = \{y_1(\tau_k), y_2(\tau_k), \ldots, y_L(\tau_k)\}$ at times τ_k: $\{t_0 \le \tau_k \le t_{final}\}$ of physical, biophysical, geophysical, or other subject of interest.
>
> After some contemplation of the forces acting on the system producing these measurements and some consideration of the environment in which the observed nonlinear dynamical system resides, one proposes some D-dimensional ($D \ge L$) dynamical equations for the state variables $\mathbf{x}(t) = \{x_1(t), x_2(t), \ldots, x_D(t)\}$ in the form of nonlinear ordinary differential equations.
>
> $$\frac{dx_a(t)}{dt} = F_a(\mathbf{x}(t), \mathbf{u}(t), \boldsymbol{\theta}); \quad a = 1, 2, \ldots, D, \tag{0.1}$$
>
> where $\boldsymbol{\theta}$ is a collection of N_p time independent parameters, $\mathbf{u}(t)$ are some time varying quantities, perhaps under the control of the observer, and $\mathbf{F}(\mathbf{x}(t), \mathbf{u}(t), \boldsymbol{\theta})$ is called the vector field of the dynamics. $\mathbf{u}(t)$ is a set of "time dependent" parameters for which no dynamical equation is usually given. It is specified outside the observed system, and it can be treated as a sequence of parameters at each time $t_n = t_0 + n\Delta t$ where the state variables $\mathbf{x}(t_n)$ are desired.
>
> How can we estimate the parameters $\boldsymbol{\theta}$ we do not know, as well as the $D - L$ state variables we do not (or cannot) observe, and the 'external' or environmental forces we may not know using the potentially sparse ($D \gg L$) and certainly noisy measurements we have acquired?

If we were able to accomplish the required estimations, we would be in a good position to ask and answer the additional, critical question:

> With estimates of *all* state variables at the end t_{final} of the observation window $[t_0, t_{final}]$ $\mathbf{x}(t_{final})$, all parameters $\boldsymbol{\theta}$, and all forces $\mathbf{u}(t \geq t_{final})$, can we predict $\mathbf{x}(t > t_{final})$ by solving the initial value problem, Eq. (0.1), with initial conditions now given at t_{final}?

As we have L observed state variables, we can check the consistency of the model output with those observations in and beyond $[t_0, t_{final}]$. In the prediction phase of data assimilation, $t \geq t_{final}$, we must have accurate estimations of the *unobserved* state variables. In this step, called 'generalization' in machine learning, we are testing both our selection of the model Eq. (0.1) and the workings of our data assimilation, information transfer, protocols.

Let's address just a few technical tidbits before proceeding:

1. If the model of the dynamics producing the data is in the form of *partial* differential equations, then one has an infinite number of degrees-of-freedom $(D \to \infty)$! So one puts the fields on a $\mathbf{x}(\mathbf{r}, t)$; $\mathbf{r} = (x, y, z)$ spatial grid (or equivalent) with $N = N_x \times N_y \times N_z$ grid points, resulting in ND ordinary differential equations of the form Eq. (0.1), which brings us back to the same discussion.

2. The 'external' quantity $\mathbf{u}(t)$ may be known, for example, if the forces driving the observed system are known. If they are not known, then in a variational treatment of the overall data assimilation problem, as discussed in Gelfand and Fomin (1963); Kirk (1970), there are equations determining $\mathbf{u}(t)$ from knowledge of the $\mathbf{y}(\tau_k)$.

3. We discuss one observation window in time $[t_0, t_{final}]$; however, if the dynamics is chaotic, there may be a need for a sequence of observation windows. The reason is that we are numerically able to estimate parameters and states to various levels of accuracy; however, such unavoidable errors are amplified by the chaotic dynamics, and one needs to observe again to put the evolving trajectory into the correct region of state space.

4. The number of measurements, called L here, made at each observation time may vary as observations are made.

5. The time Δt separating steps in the utilization of the dynamics need not be uniform across all time windows.

Who wants to do this sort of thing anyway?

The simplest answer is *everyone* working in science and technology. Many have data, many have models describing how those data emerged from observations of some dynamical system, and all need tools to transfer the information in those data to the model. Since there are unknown parameters and surely unobserved state variables, the data alone may be insufficient to provide confidence in the ability of the model to predict beyond t_{final}.

To this quite general discussion, it might be of value to discuss an example briefly. You are tasked with making measurements on a neuron isolated from its working environment and placed, *in vitro*, in a dish with the goal of developing some model dynamical equations that allow you to know (predict) with accuracy how this neuron, or its equivalent back in the working biophysical neural circuit, will respond to currents in its environment and coming through to it via synaptic or other connections to other neurons.

An Illustrative Example – Dynamical Equations for a Neuron

The biophysical equations describing the dynamics of neurons were established by work done mostly in Cambridge, UK before and after World War II. Hodgkin and Huxley (1952); Johnston and Wu (1995); Sterratt et al. (2011) were some of the researchers whose names are prominent in understanding that equations impos-ing current conservation on the ions flowing into and departing from the neuron body (soma) and equations capturing the voltage dependent permeability of the cell membrane to these ions would provide a quantitative framework for the biophysical description of the neural processes.

In the experiments they performed they considered two ion channels for Na and K ions flowing through proteins penetrating the cell membrane. They also intro-duced a 'leak' channel describing other aspects of neuron behavior. The nonlinear equations they proposed, and tested, have the form

$$C_m \frac{dV(t)}{dt} = g_{Na} m(t)^3 h(t)[E_{Na} - V(t)] + g_K n(t)^4 [E_K - V(t)]$$
$$+ g_L [E_L - V(t)] + I_{DC} + I_{app}(t). \tag{0.2}$$

The three voltage dependent 'gating' variables $a(t) = \{m(t), h(t), n(t)\}$; $0 \le a(t) \le 1$ are taken to satisfy the first order kinetics

$$\frac{da(t)}{dt} = \frac{a_0(V(t)) - a(t)}{\tau_a(V(t))}. \tag{0.3}$$

The parameters $\{g_j\}$ in the voltage equation Eq. (0.2) variables are constants while the gating variables are state variables that are voltage dependent. $a_0(V)$ and $\tau_a(V)$ are voltage dependent functions. The first is dimensionless and sets the scale for the gating variables, and the second is a voltage dependent time scale for the gating variables.

This is a $D = 4$ dimensional dynamical system. It has rich behavior (Hodgkin and Huxley (1952); Johnston and Wu (1995); Sterratt et al. (2011)). In laboratory

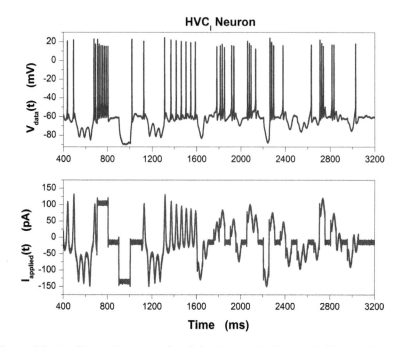

Figure 0.2 An illustrative example of the data assimilation challenge discussed in this book. This figure shows data from Daniel Margoliash of the University of Chicago and Daniel Meliza of the University of Virginia. An interneuron within the nucleus HVC of the avian brain was isolated in a glass dish in the laboratory – an *in vitro* experiment. An electrode was inserted into the body of the neuron and the applied current $I_{app}(t)$ shown in the **bottom panel** was injected into the neuron. The resulting membrane potential response, shown in the **top panel** of the display, was measured using the same electrode. From these data, one is asked to estimate all the parameters, here $N_p = 20$, and the three (unmeasured) gating variables $a(t)$ in the HH equation, Eq. (0.2).

experiments one can directly measure the cross membrane voltage $V(t)$, but no instruments are available (as of July 2020) to observe the gating variables. In the general language used here, $L = 1$, and three state variables are unobserved.

An experiment consists of selecting a current $I_{app}(t)$ (the analog of $\mathbf{u}(t)$ discussed above), and it is typically known. With only $V(t)$ observed, the challenge is to estimate all the fixed parameters in Eq. (0.2) as well as to estimate all of the $a(t)$ and all of the parameters in the $a_0(V)$ and $\tau_a(V)$ appearing in, Eq. (0.3), over $[t_0, t_{final}]$. Validation (or not) of the model associated with the observations comes from solving Eq. (0.2) for $t \geq t_{final}$, using $V(t_{final})$ and the $a(t_{final})$ as initial conditions and the estimated $\boldsymbol{\theta}$ to complete the HH equations.

If you are a 'data scientist,' there is often a directive to find a 'domain expert' with whom to work on the problem just posed. I personally encourage each reader to become a domain expert and a data scientist at the same time and not artificially divide one's self into two or more parts. The problems you wish to solve usually require both of these parts of you, and your appreciation of the data and the modeling will be increased by this strategy. If you collaborate with other domain experts and/or other data scientists to address the issues in your problem, that brings even more experience to the table.

In the instance of neurobiology, for example, I recommend the domain expert's 'manual' by Daniel Durstewitz (2017). It addresses what you need to know about neurons and then provides an easy entry into computational modeling of neurons and networks thereof. The textbook by Sterratt et al. (2011) covers less neuro-data analysis than Durstewitz (2017) but focuses more in modeling networks of neurons.

The other topic we often use as examples in this book arises in geophysics, and your road to domain expertise could be via Pedlosky (1986); Vallis (2017).

Why should we expect this will work? Because the dynamical model is non-linear in the state variables, the state variables are generically coupled together through the nonlinear model. Information is passed through the observable $V(t)$ and determines the unobserved $a(t)$ and the parameters θ, consistent with the data.

The functions $a_0(V)$ and $\tau_a(V)$ for the neurobiological problem may be estimated by looking at experimental data (Senselab-Yale (2020)) for simulations of each ion channel: Na, K, Ca, ...

This sets the challenge. We'll see how it all works, in detail, as we proceed.

A Bit More Just Before We Start Out

This book is primarily about how one effects this information transfer when the data are noisy (always) and the model has errors (also always). There are many methods for this that have been developed in various fields; and while we will note those developments, our focus here will be on the path integral formulation of the critical questions.

Why path integrals? They sound quite exotic; however, as we will see in the chapters ahead, they are formulated in a natural manner, and they are integral representations of the solutions to equations such as Eq. (0.1) when errors in the model and errors in the data are present. Such representations give us insight into global properties of the dynamical systems we will be analyzing, while the differential equations, such as Eq. (0.1), are focused on local time evolution of those equations.

We will discuss variational methods for both continuous time and discrete time. We will discuss Monte Carlo methods. We will discuss both of these using an annealing method that turns on the magnitude of the model precision, and thus the nonlinearity in an adiabatic fashion.

I will also discuss supervised Machine Learning because, as it turns out, this is mathematically equivalent to DA. So everything we say about data assimilation applies.

This is not something (yet) discussed in other books, and I hope by drawing attention to this equivalence the methods from DA will be utilized in ML, and, hopefully, *vice versa*.

Another topic I will emphasize, again not widely addressed elsewhere, is the question: How many L measurements does one require to accurately estimate the state variables $\mathbf{x}(t)$ as well as the parameters $\boldsymbol{\theta}$? As estimating each of these costs bits of information (Rissanen (1989)), accuracy must depend on how much independent data is available. Furthermore, we'll see that the answer depends on the vector field $\mathbf{F}(\mathbf{x}, \boldsymbol{\theta}, \mathbf{u})$, and the DA efficacy depends on the instabilities of the nonlinear communication protocol connecting the data, $\mathbf{y}(t)$, to the model (Kostuk (2012)).

We will select examples from geosciences and neurobiology as we proceed. As noted, the material here is often encountered by students of Physics, Chemistry, and Geophysics. It is not at all common to see it in a Neurobiology curriculum. I hope this book and the papers presenting research that precede it will become common practice in computational neuroscience as well; we'll see.

Many Thanks Are Owed

The results in this monograph cannot be claimed by me to be mine alone. My many productive interactions with former and present Physics PhD students at UCSD contributed to every word and paragraph. These women and men include Daniel Creveling, Brian Toth, Mark Kostuk, Chris Knowlton, Will Whartenby, Jack Quinn, Uriel Morone, Michael Eldridge, Jason An, Xingxin Ye, Daniel Rey, Nirag Kadakia, Sasha Shirman, and Paul Rozdeba, and I recommend their PhD dissertations to the reader. Those may be found in the University of California's archive **escholarship.org**:

> Creveling (2008); Toth (2011); Kostuk (2012); Shirman (2018); Kadakia (2017); Ye (2016); Quinn (2010); An (2019); Rey (2017); Rozdeba (2017); Knowlton (2014); Eldridge (2016); Morone (2016); Whartenby (2012).

Equally important have been my productive interactions over many years with Daniel Margoliash at the University of Chicago and Ulli Parlitz in Göttingen. This includes as well their postdoctoral fellows Daniel Meliza (now at the University of Virginia) and Arij Daou (now at the American University of Beiruit) and many PhD students, especially Jöchen Bröcker, now at the University of Reading (UK).

To this list I am pleased to add Alain Nogaret at the University of Bath (UK), Eve Armstrong (NYIT), George Michael Fuller (UCSD, Physics) and Philip Gill, Melvin Leok, Michael Holst, and Randy Bank (UCSD, Mathematics), Gert Cauwenberghs, Gabriel Silva (UCSD, Bioengineering) and Tim Gentner (UCSD, Psychology and Neurobiology), Michael Long (NYU, Neuroscience), Bruce Cournelle and Art Miller (UCSD, Scripps Institution of Oceanography), and Alexandre Chorin (UCB, Mathematics).

Finally, I am indebted to my old and reliable friends Robert Sugar (University of California Santa Barbara, Physics) and Jerry Marsden (UC Berkeley and Caltech) who have tolerated and often dissipated my confusion about many topics in statistical Physics for many years. Marsden's youthful death did not impede the importance of his contributions to the ideas we discuss here.

My wife has always encouraged my efforts by noting my results are "obvious" and looking forward to even more.

1

A Data Assimilation Reminder

1.1 Recalling the Basic Idea of Statistical Data Assimilation

This is an expansion of the discussion of data assimilation in Abarbanel (2013). There we further developed a path integral (Abarbanel (2009); Cox (1964)) approach to the subject of data assimilation, which was illustrated by examples from nonlinear electrical circuits, chaotic fluid dynamics, and laboratory neurobiological experiments. The neurobiological experiments were performed in the laboratory of Daniel Margoliash and his students and postdoctoral fellows at the University of Chicago. The designation 'data assimilation' made its first appearance, to my knowledge, within the community of meteorologists working on numerical weather prediction (Anthes (1974); Ghil and Malanotte-Rizzoli (1991); Pires et al. (1996); Kalnay (2003); Lorenc and Payne (2007); Evensen (2009); Reich and Cotter (2015) and climate modeling. It was understood, probably from the outset of these endeavors in the 1950s, that one required knowledge of the state of the earth system, the atmosphere, and the ocean, at some time t_{final} in order to use the equations of fluid dynamics, so-called General Circulation Models, to predict forward in time for $t \geq t_{final}$. The prediction is the validation (or not) of the model proposed to represent the source of the data. Just "fitting" the observed data is a consistency check on the information transfer methods, but one must do more.

The reason one needs prediction, known in Machine Learning as 'generalization,' is that many, usually most, of the model state variables are **unobserved**. One requires them, however, to predict forward for $t \geq t_{final}$, and, as they are unobserved or even unobservable, one cannot measure them directly, so their role in prediction is the only method for probing the accuracy with which we have estimated them via data assimilation.

One problem was, and remains, that we have only an approximate idea what the state of the earth system is at **any** time with enough accuracy and spatial coverage to have confidence in those predictions. The situation simply got worse when, in 1963, the seminal paper of Ed Lorenz (1963) showed that the intrinsic instabilities of many nonlinear systems, certainly including fluid dynamics on a spatial grid, amplified small errors in initial conditions as well as errors in fixed physical parameters and would lead to exponential growth in those errors. It was hoped that with 'enough' measurements of the state space of the earth system one could pass 'enough' information to the physical dynamical models to rectify the discouraging prospect one faced.

In the literature that I have reviewed (Ghil and Malanotte-Rizzoli (1991); Pires et al. (1996); Kalnay (2003); Lorenc and Payne (2007); Evensen (2009); Reich and Cotter (2015)) the important question of how many measurements are actually required to make accurate predictions is not addressed. We will address this question.

This transfer of information in measured state variables to 'complete' physical models by estimating all the **unmeasured** state variables *and* all the unknown or poorly known time independent parameters of the model acquired the name 'data assimilation.'

In the discussions here and earlier (Abarbanel (2013)), we have called this process of information transfer *Statistical Data Assimilation; SDA* to emphasize its generality across many disciplines where the nonlinear Physics of the problems at hand are important and the value of viewing it as part of considerations in Statistical Physics as developed since the nineteenth century.

The statistical part of the designation SDA comes from unavoidable noise in the measurements and errors in the models. How one represents errors in a model is not at all a settled subject, but some statement must be made, and, naturally, we will do so.

In the process of SDA we require three critical ingredients to transfer information in observations to properties of models proposed to represent the source of that information.

- We should have well curated data. 'Curation' means we should understand not just the binary or ascii numbers presented as 'data,' but we also should understand the instruments used to collect the data. We should have knowledge of the calibration of these instruments and, if at all possible, we should have knowledge of the errors, whatever their source, in these data. We should also know the statistical distribution of these errors. The information here, in addition to the raw ascii numbers, is often called 'metadata', and it is not always available. A good experiment will provide the data and the metadata; be sure to ask for it.

- We will have a **model** of the processes that produced the well curated data we receive. In a physical or biophysical setting, we may have some guidance for the construction of such a model, or we may be proposing a model whose consistency with the data we wish to examine and whose validity we wish to establish.
- We must have a method for transferring the information from the data to the model: this comprises estimating all model variables that are **unobserved** in the measurement processes as well as estimating the value of time independent parameters in the model. Some parameters are known, perhaps from other observations, but we wish to estimate all those that we do not know.

The topics in this book are primarily focused on the third item. We do not provide guidance here to propose and design experiments, in the laboratory or in the field, for every domain of science one might wish to address. We have actually worked with members of the Margoliash neurobiology laboratory at the University of Chicago to design laboratory experiments that produce excellent predictions; examples of this are discussed throughout this book.

Similarly, we do not propose to discuss specific models for all areas of scientific inquiry. We do hope to convey the overall principles and issues in SDA and to illustrate these with examples.

The third item provides a methodology to transfer the information to *a priori* ill-informed properties of the model; in particular, one wishes to estimate time independent parameters and unobserved states. The thrust of this volume is **not** to provide a model, typically in the form of a differential equation for the state variables involved in the processes generating the data. Providing a model for use in SDA rests on the experience and insight of the user, and it is not a button to push called *Give me a model, please* in some package of algorithms. We will formulate a general framework for the operations utilized to transfer the information in the data to properties of the model, once the data are collected and the models are formulated.

Models are, in a sense, the 'art' of data assimilation. It is in this that the skill of the scientist is displayed. It is a matter of insight and some experience to formulate models. The context of SDA is, first of all, to establish whether a proposed model is consistent with the data. Only consistency is possible as we have no knowledge of the unobserved state variables. So while we should be using measurements $\mathbf{y}(\tau_k)$ in the observation window we cannot check if the unobserved state variables are correct, because, well, we do not know them; similarly with the unknown time dependent parameters.

All state variables and all parameters come into play when we want to validate (or not) the model by using it, deterministically or statistically, to predict the behavior of the observed system after $[t_0, t_{final}]$.

1.2 What Is in the Following Chapters?

We ended Abarbanel (2013) with a discussion entitled "Unfinished Business," and we will spare the diligent reader having to go over our state of ignorance those few years ago by recalling what was "unfinished."

We then will recall our notation and formulation of SDA in a section called "Remembrance of Things Path" and identify the items we'll consider in this collection of writings.

This seems a good place to apologize to Proust (Proust (1913)) and all of his dedicated readers for our choice of a pun here. How could one help but do it?

Picking up from the items of *Unfinished Business* we will address how to use the information in the waveforms of the data as a function of time. Then we will apply the ideas to a useful instructional model and to an interesting geophysics problem.

- SDA Variational Principles; Euler-Lagrange Equations for SDA Variational Calculations; Using Waveform Information; Lorenz96 Examples; Lagrangian Drifters and Shallow Water Flows
- Annealing in the Model Precision R_f
- Symplectic Integration and SDA Variational Principles; "Fokker-Planck Equation" for the SDA Standard Model
- Monte Carlo Methods; Metropolis-Hastings – Random Proposals; Hamiltonian Monte Carlo Methods – Structured Proposals or Symplectic Proposals
- SDA and Its Equivalence to Supervised Machine Learning; $\langle A(\mathbf{X}) \rangle = \langle - \log P (\mathbf{X}|\mathbf{Y}) \rangle$.

Many of these items were not *unfinished* in 2012; indeed, they were not even known as items to require attention at that time.

2

Remembrance of Things Path

To set a common stage for what follows, we recall the formulation in Abarbanel (2013) and also slightly generalize some basic notation for both Statistical Data Assimilation (SDA) as well as our discussion of Supervised Machine Learning.

In SDA we have a segment of time starting at $t = t_0$ and running to a final time $t = t_{final}$. This interval may be repeated as needed for a specific application, but for now we have in mind just one such interval. The interval is divided into N smaller time segments of size Δt, and these times are $t_n = t_0 + n\Delta t$; $n = 0, 1, 2, \ldots, N$; $t_{final} = t_0 + N\Delta t = t_N$.

In this time interval $[t_0, t_{final}]$, we encounter two kinds of operations:

- We make observations at F distinct times $t = \tau_k$; $k = 1, 2, \ldots, F$ where each τ_k is t_0 plus a multiple of Δt: $\tau_k = t_0 + n_k\Delta t$; n_k is an integer. $t_0 \leq \tau_k \leq t_{final}$. There are L observations at each observation time τ_k. One is not required to make τ_k a multiple of Δt, but I am trying to prevent a notation tsunami so we can concentrate on the underlying questions in this book.
- We have a D-dimensional model with state variables $\mathbf{x}(t) = \{x_1(t), x_2(t), \ldots, x_D(t)\}$. $D \geq L$ and typically $D \gg L$. The model is moved along in time by steps of length Δt through the dynamical rule $\mathbf{x}(t_{m+1}) = \mathbf{x}(m + 1) = \mathbf{f}(\mathbf{x}(t_m), \boldsymbol{\theta}) = \mathbf{f}(\mathbf{x}(m), \boldsymbol{\theta})$; $m = 0, 1, 2, \ldots, N - 1$. $\boldsymbol{\theta}$ is a collection of N_p time independent parameters.

When the argument of the model state t_m reaches a time τ_k, L measurements are made, and we indicate the observations at these times as an L-dimensional vector: $\mathbf{y}(\tau_k) = \{y_1(\tau_k), y_2(\tau_k), \ldots, y_L(\tau_k)\}$.

This movement through the observation or measurement window is depicted in Fig. 2.1.

Another shorthand notation will be useful. We will call the full collection of state variables from time t_0 to time $t_{final} = t_N$ the D(N+1) dimensional **path** vector **X**.

Data Assimilation in a time window $[t_0, t_{\text{final}}]$

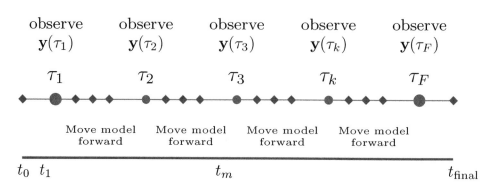

Figure 2.1 Timeline for moving through an observation window $[t_0, t_{final}]$ in SDA. Measurements are made at times $\tau_k = t_0 + n_k \Delta t$; $k = 1, 2, \ldots, F$. n_k is an integer. State variables $\mathbf{x}(t_n)$ are stepped forward by the nonlinear rule $\mathbf{x}(t_{n+1}) = \mathbf{f}(\mathbf{x}(t_n), \boldsymbol{\theta}) + \text{noise}$; $t_n = t_0 + n\Delta t$.

We denote the D(n+1) dimensional path up until time t_n; $n = 0, 1, 2, \ldots, N$ as $\mathbf{X}(n) = \{\mathbf{x}(0), \mathbf{x}(1), \ldots, \mathbf{x}(n)\}$; $\mathbf{X}(N) = \mathbf{X}$.

When we wish to, we will add the parameters $\boldsymbol{\theta}$ to the path in order to put on a common footing the estimation of both the state variables of our proposed model and the parameters in the model. If there are N_p parameters in $\boldsymbol{\theta}$, the length of the full path over the interval $[t_0, t_{final}]$ is $\mathcal{D} = (D(N+1) + N_p)$.

We will call the collection of observations through the entire observation window $[t_0, t_{final}]$ $\mathbf{Y}$. $\mathbf{Y}$ is an LF dimensional vector. The set of observations up until time t_n we call $\mathbf{Y}(n) = \{\mathbf{y}(0), \mathbf{y}(1), \mathbf{y}(2), \ldots \mathbf{y}(n)\}$. If no observation is made at a time $\tau_j \le t_n$, that vector τ_j is absent in $\mathbf{Y}(n)$.

2.1 Recursion Relation along the Path X

We wish to characterize the knowledge we have of the state variables up to time t_n, $\mathbf{X}(n)$, given all the measurements up to t_n, $\mathbf{Y}(n)$. The measurements are always noisy, and the dynamical rule $\mathbf{x}(t_{n+1}) = \mathbf{x}(n+1) = \mathbf{f}(\mathbf{x}(t_n), \boldsymbol{\theta}) = \mathbf{f}(\mathbf{x}(n), \boldsymbol{\theta})$ also has errors. We express this uncertainty in terms of a conditional probability distribution $P(\mathbf{X}(n)|\mathbf{Y}(n))$.

There is a recursion relation between $P(\mathbf{X}(n+1)|\mathbf{Y}(n+1))$ and $P(\mathbf{X}(n)|\mathbf{Y}(n))$, and this is fundamental to the path integral for addressing data assimilation problems. In addressing the path integral formulation of machine learning, the time label changes $\rightarrow$ the layer variable. This will be explored in detail in Chapter 8.

Here is a (little bit lengthy) derivation of the recursion relation:

$$\mathbf{P(X(n+1)|Y(n+1))} = \frac{P(\mathbf{X}(n+1), \mathbf{Y}(n+1))}{P(y(n+1), \mathbf{Y}(n))}$$

$$= \left\{ \frac{P(y(n+1), x(n+1), \mathbf{X}(n), \mathbf{Y}(n))}{P(y(n+1), \mathbf{Y}(n)) \, P(x(n+1), \mathbf{X}(n), \mathbf{Y}(n))} \right\} \bullet$$
$$P(x(n+1)|\mathbf{X}(n), \mathbf{Y}(n)) \bullet P(\mathbf{X}(n), \mathbf{Y}(n))$$

$$= \left\{ \frac{P(y(n+1), x(n+1), \mathbf{X}(n)|\mathbf{Y}(n))}{P(y(n+1)|\mathbf{Y}(n)) \, P(x(n+1), \mathbf{X}(n)|\mathbf{Y}(n))} \right\} \bullet$$
$$P(x(n+1)|\mathbf{X}(n), \mathbf{Y}(n)) \bullet P(\mathbf{X}(n)|\mathbf{Y}(n))$$

$$= \exp[CMI(y(n+1), x(n+1), \mathbf{X}(n)|\mathbf{Y}(n))] \bullet$$
$$P(x(n+1)|\mathbf{X}(n), \mathbf{Y}(n)) \bullet P(\mathbf{X}(n)|\mathbf{Y}(n))$$

$$= \left\{ \frac{P(y(n+1), x(n+1), \mathbf{X}(n)|\mathbf{Y}(n))}{P(y(n+1)|\mathbf{Y}(n)) \, P(x(n+1), \mathbf{X}(n)|\mathbf{Y}(n))} \right\} \bullet$$
$$P(x(n+1)|x(n)) \bullet \mathbf{P(X(n)|Y(n))}$$

$$= \underbrace{P(y(n+1)|x(n+1), \mathbf{X}(n), \mathbf{Y}(n))}_{\textbf{Information Transfer}} \bullet \underbrace{P(x(n+1)|x(n))}_{\textbf{State Transition Probability}} \bullet$$

$$\mathbf{P(X(n)|Y(n))} \bullet \textbf{terms independent of X} \tag{2.1}$$

where Shannon's conditional mutual information (CMI) (Fano (1961)) is identified:

$$CMI(a, b|c) = \log\left[\frac{P(a, b|c)}{P(a|c)\, P(b|c)}\right]. \tag{2.2}$$

Here $a = y(n+1)$, $b = x(n+1)$, $\mathbf{X}(n)$, and $c = \mathbf{Y}(n)$.

The appearance of the CMI reminds us that the transfer of information acts as a 'potential' whose gradient represents a force based on information flow from the data to our model in a unidirectional manner: from the data, **Y**, to the model we adopt for the time evolution of the state variables. This term, which we call the "measurement error," depends on the measurements at $y(t_n)$ and observations in the past $t_0, t_1, t_2, \ldots, t_{n-1}$, as follows from the derivation of the recursion relation.

We emphasize this aspect of the recursion relation not simply to be picky about having a correct formula, but to note that in this **information transfer** term, we also see how the observations $y(n+1)$ may be dependent on the state $\mathbf{X}(n+1)$ and earlier observations $\mathbf{Y}(n)$. This guides us to the manner in which some of the

"Unfinished Business" taken forward from Abarbanel (2013) may be addressed, and we pick this up in Chapter 4.

In our derivation of the recursion relation we have explicitly used the Markov property of the dynamics which states that the model state variable at time t_{n+1}, $\mathbf{x}(n+1)$, depends solely on the state at the previous time, $\mathbf{x}(t_n) = \mathbf{x}(n)$:

$$P(x(n+1)|\mathbf{X}(n), \mathbf{Y}(n)) = P(x(n+1)|x(n)). \tag{2.3}$$

Any differential equation in continuous time or map in discrete time proposed as the source of the observations, in problems from physical sciences, satisfies this criterion. So this is not a hardship in our discussion.

In addition to this Markov property, we used the definition of *conditional probabilities*, namely $P(a|b) = P(a, b)/P(b)$. This is also known widely as Bayes' theorem.

These two items are all we require to establish the recursion relation, Eq. (2.1), so we see that it is widely applicable to the problems we will address.

2.2 The 'Action' $A(\mathbf{X}) = -\log\left[P(\mathbf{X}|\mathbf{Y})\right]$

If we now iterate the recursion relation, starting from $t = t_{final} = t_N$ and proceeding back to t_0, we find

$$P(\mathbf{X}|\mathbf{Y}) = \prod_{n=1}^{N} P(\mathbf{y}(n)|\mathbf{x}(n), \mathbf{X}(n-1), \mathbf{Y}(n-1)) \bullet$$

$$\prod_{n=0}^{N-1} P(\mathbf{x}(n+1)|\mathbf{x}(n)) P(\mathbf{x}(0)) \bullet [\textbf{terms independent of X}]. \tag{2.4}$$

Once the (noisy) data are provided to us, we set them aside in a data library until we require them. We also desire 'metadata' comprising characteristics of the instruments acquiring the data.

We cannot visualize the detail of the very high dimensional conditional probability distribution $P(\mathbf{X}|\mathbf{Y})$, so we turn to the evaluation of conditional expected values of functions of $\mathbf{X}$, $G(\mathbf{X})$. Interesting $G(\mathbf{X})$ are selected by the user.

Of certain importance is the expected path through $[t_0, t_{final}]$. For this we would choose $G(\mathbf{X}) = \mathbf{X}$. We might well also be interested in moments about this expected path. There are many other choices of $G(\mathbf{X})$, and it is the interest of the user that will dictate one's choice.

These expected values have the form

$$E[G(\mathbf{X})|\mathbf{Y}] = \langle G(\mathbf{X}) \rangle = \frac{\int d\mathbf{X}\, G(\mathbf{X}) \exp[-A(\mathbf{X})]}{\int d\mathbf{X}\, \exp[-A(\mathbf{X})]}, \tag{2.5}$$

in which the *action* $A(\mathbf{X}) = -\log[P(\mathbf{X}|\mathbf{Y})]$ is introduced, and terms independent of $\mathbf{X}$ have canceled between the numerator and the denominator. In the action the dependence on the observations $\mathbf{Y}$ is not shown, but that is only for some brevity of notation. Note that $A(\mathbf{X}) \geq 0$.

As we no longer need to consider the terms independent of $\mathbf{X}$, the effective action in Eq. (2.5) has the following components

$$A(\mathbf{X}) = \underbrace{-\sum_{n=1}^{N} \log[P(\mathbf{y}(n)|\mathbf{x}(n), \mathbf{X}(n-1), \mathbf{Y}(n-1))]}_{\textbf{Information Transfer}}$$

$$\underbrace{-\sum_{n=0}^{N-1} \log[P(\mathbf{x}(n+1)|\mathbf{x}(n))]}_{\textbf{Transition Probability}}$$

$$\underbrace{-\qquad \log P(\mathbf{x}(0))}_{\textbf{Probability of Model States at } t_0} \qquad (2.6)$$

2.3 Multiple Measurement Windows in Time

There is no barrier to introducing multiple measurement windows $[t_0, t_{final}]$, though, in principle, one might be sufficient to the task of completing the model. Three circumstances, at least, might require this:

- one or more of the 'fixed' parameters $\boldsymbol{\theta}$ may vary on a longer time scale than $|t_{final} - t_0|$, or
- the dynamics of the model, absent observations, may be chaotic, whereby the evolution of the states $\mathbf{x}(t)$ can lose track of where in phase space it should reside, and a reminder in the form of additional measurements is called for, or
- as the integrals associated with Eq. (2.5) must be evaluated numerically, the errors so introduced may cause a deviation from the correct answer indicated, within measurement errors, by the observations.

In the case of numerical weather prediction, for example, all three of these may be present.

2.4 The Standard Model for SDA

Starting with the third term in the action, $P(\mathbf{x}(0))$, we recall that in general we may not know much, if anything, about this. If there has been an observation window before the one opening at t_0, then the output from the earlier window could be useful here. If not, then, if the overall interval $t_{final} - t_0$ is long enough, then it may not much matter what precise form for $P(\mathbf{x}(0))$ is selected. In less fancy language, the nonlinear problem of propagating $P(\mathbf{x}(0))$ forward through the dynamics, $\mathbf{x}(n+1) = \mathbf{f}(\mathbf{x}(n), \boldsymbol{\theta}) + \text{noise}$, and through alterations of $P(\mathbf{X}|\mathbf{Y})$ through measurements 'forgets' its initial condition if $[t_0, t_{final}]$ is long enough.

On the second term: if the transition from $\mathbf{x}(n) \rightarrow \mathbf{x}(n+1)$ is precise and error free, $P(\mathbf{x}(n+1)|\mathbf{x}(n))$ is equal to $\delta^D(\mathbf{x}(n+1) - \mathbf{f}(\mathbf{x}(n), \boldsymbol{\theta}))$. If the transition probability $P(\mathbf{x}(n+1)|\mathbf{x}(n))$ is not error free, we must select a way to 'broaden' the delta function, and we have many choices.

A convenient choice is a Gaussian approximation to $\delta^D(\mathbf{x}(n+1) - \mathbf{f}(\mathbf{x}(n), \boldsymbol{\theta}))$:

$$\delta^D(\mathbf{x}(n+1) - \mathbf{f}(\mathbf{x}(n), \boldsymbol{\theta})) \rightarrow \text{[Normalization]} \exp\left[-\frac{R_f}{2}\right.$$
$$\left. \sum_{a=1}^{D}(x_a(n+1) - f_a(\mathbf{x}(n), \boldsymbol{\theta}))^2\right], \qquad (2.7)$$

which becomes the required delta function as the *precision* $R_f \rightarrow \infty$. R_f is a scalar here, but could be a $D \times D$ matrix.

Turning to the first term, we again have many choices. If we ignore the dependence of $P(y(n)|x(n), \mathbf{X}(n-1), \mathbf{Y}(n-1))$ on $\mathbf{X}(n-1)$ and $\mathbf{Y}(n-1)$, this means we are also assuming that measurements at times t_n and times $t_k \neq t_n$ are statistically independent. This is *inconsistent* with a dynamical model $\mathbf{x}(n+1) = \mathbf{f}(\mathbf{x}(n), \boldsymbol{\theta})$ which enters in the second, model error term of the action. We will address this inconsistency in Chapters 4 and 5, where we discuss ideas on how to recover the information we have discarded.

Further, if we assume that the noise in the measurement $\mathbf{y}(n)$ is Gaussian with zero mean and precision $R_m(t)$, then

$$P(\mathbf{y}(n)|\mathbf{x}(n), \mathbf{X}(n-1), \mathbf{Y}(n-1)) \rightarrow \text{[Normalization]} \exp\left[-\frac{R_m(t)}{2}\right.$$
$$\left. \sum_{l=1}^{L}(x_l(n) - y_l(n))^2\right], \qquad (2.8)$$

and we require $R_m(t)$ to vanish at all times $t \neq \tau_k$, we arrive at what we call the *standard model* for SDA:

$$A(\mathbf{X}) = \sum_{k=1}^{F} \frac{R_m(\tau_k)}{2} \sum_{l=1}^{L}(x_l(\tau_k) - y_l(\tau_k))^2$$
$$+ \sum_{n=0}^{N-1} \frac{R_f}{2} \sum_{a=1}^{D}(x_a(n+1) - f_a(\mathbf{x}(n), \boldsymbol{\theta}))^2$$
$$- \log[P(\mathbf{x}(0))]. \qquad (2.9)$$

When we discuss the relationship between SDA and supervised machine learning in a later chapter, this standard model will also play a role.

We adopt the convention that we know essentially nothing about $P(\mathbf{x}(0))$ and take it to be uniformly distributed over the observed dynamical range of $\mathbf{x}(0)$. The $P(\mathbf{x}(0))$ term is then a constant where it is not zero, and it cancels out, between the numerator and denominator, in the expected value of functions of the state variables $\langle G(\mathbf{X}) \rangle$; see Eq. (2.5).

If it happens we know something about $P(\mathbf{x}(0))$, use it.

The standard model action has the form

$$A_{SM}(\mathbf{X}) = \sum_{k=1}^{F} \frac{R_m(\tau_k)}{2} \sum_{l=1}^{L} (x_l(\tau_k) - y_l(\tau_k))^2$$
$$+ \sum_{n=0}^{N-1} \sum_{a=1}^{D} \frac{R_f^{(x_a)}}{2} (x_a(n+1) - f_a(\mathbf{x}(n), \boldsymbol{\theta}))^2. \quad (2.10)$$

If $\mathbf{f}(\mathbf{x}, \boldsymbol{\theta})$ is nonlinear in $\mathbf{x}$, the action is not quadratic in the components of $\mathbf{X}$, so $P(\mathbf{X}|\mathbf{Y})$ is not Gaussian in $\mathbf{X}$. Were it Gaussian, the expected value integral, Eq. (2.5), could be evaluated analytically, and we would be done.

2.4.1 Complicating the Standard Model

There are many ways to complicate the Standard Model. There is no *a priori* reason why the measurement errors or the model errors should have Gaussian distributions. There is no *a priori* reason why $R_m(t)$ should be a scalar instead of an $L \times L$ matrix. There is no *a priori* reason why R_f should be a scalar instead of a $D \times D$ matrix.

Most of the changes one might make are not critical. One is.

In the measurement error term there is no relation between the $\mathbf{y}(t)$ and the $\mathbf{y}(t')$ at any other time; yet we ask for the measured variables $y_l(t)$ to match their counterparts $x_l(t)$ which are correlated through $\mathbf{x}(n) \to \mathbf{x}(n+1) = \mathbf{f}(\mathbf{x}(n), \boldsymbol{\theta})$. This means the *waveforms* $\mathbf{y}(t)$ should be present in the action, not just the point measurements at those times when the model state variables match the measurements $\mathbf{y}(t)$. The general formulation allows this, but our simplification appears not to do so. We will address this later (Chapter 5).

2.5 The Standard Model Action for the Hodgkin-Huxley NaKL Model

We can illustrate construction of the Standard Model Action by looking back to the Hodgkin-Huxley neuron model embodied in Eq. (2.11) and Eq. (0.3).

The model has four state variables $\{V(t), m(t), h(t), n(t)\}$. The first of these is the membrane voltage, and it is observable. $V(t)$ has a dynamical range from about -80 mV to about $+40$ mV. The final three are the voltage dependent opening probability for Na ion channels $\{m(t), h(t)\}$ and the voltage dependent opening probability for the K channel $\{n(t)\}$. The gating variables $a(t) = \{m(t), h(t), n(t)\}$ all have a range $0 \leq a(t) \leq 1$, and they are unobservable in the experiments (at this time, 2020).

The time evolution of the states is governed by the following differential equations for $V(t)$ and first order kinetics for $a(t)$:

$$C_m \frac{dV(t)}{dt} = I_{\text{inj}}(t) + g_{\text{Na}} \, m(t)^3 \, h(t) \left(E_{\text{Na}} - V(t) \right)$$
$$+ g_{\text{K}} \, n(t)^4 \left(E_{\text{K}} - V(t) \right) + g_{\text{L}} \left(E_{\text{L}} - V(t) \right), \qquad (2.11)$$

where $I_{\text{inj}}(t)$ is an external stimulus driving the nonlinear oscillator (the neuron).

The $a(t)$ satisfy the first order kinetics:

$$\frac{da(t)}{dt} = \frac{a_0\big(V(t)\big) - a(t)}{\tau_a\big(V(t)\big)},$$

and

$$a_0\big(V(t)\big) = \frac{1}{2} + \frac{1}{2} \tanh \left(\frac{V(t) - V_a}{\Delta V_a} \right),$$

$$\tau_a\big(V(t)\big) = \tau_{a0} + \tau_{a1} \left[1 - \tanh^2 \left(\frac{V(t) - V_a}{\Delta V_a} \right) \right]. \qquad (2.12)$$

The Standard Model Action has the structure

$$A(\mathbf{X}) = \frac{R_m}{2F} \sum_{k=1}^{F} (x_1(\tau_k) - y(\tau_k))^2 + \sum_{a=1}^{4} \sum_{k=0}^{N-1} \frac{R_f^{(x_a)}}{2N} [x_a(k+1) - f_a(x(k), \boldsymbol{\theta})]^2,$$

$$A(\mathbf{X}) = \frac{R_m}{2F} \sum_{k=1}^{F} (V(\tau_k) - V_{obs}(\tau_k))^2$$

$$+ \sum_{a=1}^{4} \sum_{k=0}^{N-1} \frac{R_f^{(x_a)}}{2N} [x_a(k+1) - f_a(x(k), \boldsymbol{\theta})]^2. \qquad (2.13)$$

In the measurement term of the action, only the observed voltage $y(t) = V_{obs}(\tau_k)$ and the model output voltage $x_1(t) = V(t)$ appear.

However, in the model error part of the action we see a weight $R_f^{(x_a)}$ for $a = \{V(t), m(t), h(t), n(t)\}$. By choosing $R_f^{(V)} \approx 10^{-1}$ and each of $\{R_f^{(m)}, R_f^{(h)}, R_f^{(n)}\} \approx 10^3$, the terms in the model error entry to the action are of approximately equal magnitude. Making such a selection is not strictly necessary, however, it makes for useful numerical stability and accuracy.

2.6 Twin Experiments

How is one to evaluate the reliability of the information transfer methods? We highly recommend the use of numerical "twin experiments" for this evaluation. These proceed as follows:

- using the proposed model with some choice of time independent parameters and some choice of initial conditions, integrate the model forward from time $t = t_0$ to a later time $t = t_{final}$. Use your preferred differential equation integrator (Press et al. (2007)).
- add noise of your choice to all state variables.
- these are your data, and as you have generated them from your known model, you know everything there is to know about these data and the model of the processes that generated them. Put them into a safe library and do not change them again.
- select as your observations $L = 1, 2, \ldots$ noisy data time series from the $D \geq L$ possible model state variables. These should be in an interval $t_0 \leq t_F \leq t_{final}$.
- using these $L \leq D$ observed quantities, employ your choice of data assimilation protocol to estimate the **unobserved** state variables ($D - L$ of them) and to estimate all unknown time independent parameters $\boldsymbol{\theta}$ in your selected model. This completes your model and provides you with an estimation of the full state of the model at time $t = t_F$.
- using your estimates of the model state at t_F and the parameters, predict the model state for $t_F \leq t \leq t_{final}$ with no further information being transferred from your 'observed' data to the model.
- compare your model states to their observed values in $t_F \leq t \leq t_{final}$ to validate, or not, the SDA procedure used.

Performing a twin experiment allows you to test your SDA protocol, and it may help you design a laboratory or field experiment that accurately yields values, with associated errors, of the $\boldsymbol{\theta}$ and the $D - L$ unobserved state variables at t_{final} (Toth et al. (2011); Kostuk et al. (2012)). We will see this working as we proceed.

3

SDA Variational Principles

Euler-Lagrange Equations and Hamiltonian Formulation

3.1 Estimating Expected Value Integrals

It is implausible that we can visualize the full conditional probability distribution $P(\mathbf{X}|\mathbf{Y})$, as it is very high dimensional. The dimension of the full path space $\mathbf{X}$ is $\mathcal{D} = (D(N+1)+N_p)$. D is the number of degrees of freedom of the state variables $\mathbf{x}(t)$; this can easily be hundreds or thousands or more. N + 1 is the number of time steps of size Δt in the observation window. N_p is the number of time independent parameters $\boldsymbol{\theta}$ in the model.

One of this book's main themes is to follow the approach in Statistical Physics, and examine the expected values along the path. These expected values are given by

$$E[G(\mathbf{X})|\mathbf{Y}] = \langle G(\mathbf{X}) \rangle = \frac{\int d\mathbf{X}\, G(\mathbf{X}) \exp[-A(\mathbf{X})]}{\int d\mathbf{X}\, \exp[-A(\mathbf{X})]},$$

$$\langle G(\mathbf{X}) \rangle = \int d\mathbf{X}\, G(\mathbf{X})\, P(\mathbf{X}|\mathbf{Y})$$

$$\text{where} \quad A(\mathbf{X}) = -\log[P(\mathbf{X}|\mathbf{Y})]. \tag{3.1}$$

Clearly there are many choices for $G(\mathbf{X})$, and some are more interesting than others. It is certain that one wants to estimate the expected path itself. In this case we choose $G(\mathbf{X}) = X_\alpha$, and we label this expected value $\langle X_\alpha \rangle = \bar{X}_\alpha; \alpha = 0, 1, 2, \ldots, \mathcal{D}$.

It is also quite natural to look at the variations about the expected path $\bar{X}_\alpha$, in which case we choose $G(\mathbf{X}) = (X_\alpha - \bar{X}_\alpha)(X_\beta - \bar{X}_\beta)$. This statistic gives us a sense of the breadth of the distribution $P(\mathbf{X}|\mathbf{Y})$ along various directions in path space X_α. In particular applications other choices for $G(\mathbf{X})$ may be important. Another interesting choice for $G(\mathbf{X})$ is $G(\mathbf{X}) = A(\mathbf{X})$, which gives us the entropy:

$$\langle A(\mathbf{X}) \rangle = \int d\mathbf{X}\,[-P(\mathbf{X})]\,\log[P(\mathbf{X})]. \tag{3.2}$$

The integral Eq. (3.1) is generally intractable in terms of simple functions.

However, one case has been used over and over again, *regardless of its validity*, namely the use of a quadratic approximation to the action reached by Taylor expanding about some path $\mathbf{X}^1$, and then terminating the expansion at the quadratic term. We call that action $A_2(\mathbf{X})$:

$$A_2(\mathbf{X}) = A(\mathbf{X}^1) + (X - X^1)_\alpha \mathbf{B}_\alpha(\mathbf{X}^1)$$
$$+ \frac{1}{2}(X - X^1)_\alpha (X - X^1)_\beta \mathbf{M}_{\alpha\beta}(\mathbf{X}^1). \tag{3.3}$$

The Gaussian integral in the denominator of Eq. (3.1) can be evaluated exactly

$$\int d^{\mathcal{D}}\mathbf{X}\exp[-A_2(\mathbf{X})] = e^{-A(\mathbf{X}^1)} \frac{(2\pi)^{\mathcal{D}/2}}{\sqrt{\det \mathbf{M}}} \exp[-(\mathbf{B}\mathbf{M}^{-1}\mathbf{B})/2], \tag{3.4}$$

where

$$\mathbf{M}_{\alpha\beta} = \frac{\partial^2 A(\mathbf{X})}{\partial X_\alpha \partial X_\beta}\bigg|_{\mathbf{X}^1} \text{ and } \mathbf{B} = \frac{\partial A(\mathbf{X})}{\partial X_\alpha}\bigg|_{\mathbf{X}^1}. \tag{3.5}$$

Similarly, expanding $G(\mathbf{X})$ in a Taylor series around $\mathbf{X}^0$ permits each term in the numerator of the integral for $\langle G(\mathbf{X})\rangle$ to be evaluated analytically. In the general expression for the expected value of $G(\mathbf{X})$ this approximation becomes relevant when we make the dynamical rule moving the state $\mathbf{x}(n) \rightarrow \mathbf{x}(n+1)$ *linear*: $x_a(n+1) = \sum_{b=1}^{D} C_{ab}x_b(n)$. C_{ab} is a time independent matrix.

This is almost always uninteresting, as most realistic problems in science and technology are nonlinear.

3.2 Laplace's Method for Estimating Expected Value Integrals

The approximation method for estimating the integral Eq. (3.1) proposed by P. S. Laplace some time ago (almost 250 years ago, actually) (Laplace (1774, 1986)) has been used in a wide variety of scientific fields from high energy physics, where it is known as perturbation theory, to data assimilation. The idea is to seek paths $\mathbf{X}^q$ where the action $A(\mathbf{X})$ varies 'slowly' and expand the action about such a path. The method takes a variety of names here and there in its application to scientific questions: stationary phase method, saddle point method, and maybe others.

One proceeds by looking for paths $\mathbf{X}^q$; $q = 1, 2, \ldots q_{max}$ where the action is stationary and a minimum. These paths correspond to the peaks of $P(\mathbf{X}|\mathbf{Y})$, and their contribution to the expected value integrals, Eq. (3.1), is anticipated to be dominant. In regions of path space where the action varies rapidly, we can expect the contribution to the expected value integral to be small. Also we anticipate that there may be many minima, $q > 1$ as the action is nonlinear in the components of the path $\mathbf{X}$.

This approximation requires that at $\mathbf{X} = \mathbf{X}^q$ the first derivative with respect to $\mathbf{X}$ be zero, and the second derivative, a matrix in path space, be positive definite. This means

$$\mathbf{B}_\alpha(\mathbf{X}^q) = \left.\frac{\partial A(\mathbf{X})}{\partial \mathbf{X}_\alpha}\right|_{\mathbf{X}^q} = 0, \ q = 1, 2, \ldots$$

and

$$\mathbf{M}_{\alpha\beta}(\mathbf{X}^q) = \left.\frac{\partial^2 A(\mathbf{X})}{\partial \mathbf{X}_\alpha \, \partial \mathbf{X}_\beta}\right|_{\mathbf{X}^q} \tag{3.6}$$

has only positive eigenvalues.

One may expand the action about the paths $\mathbf{X}^q$:

$$A(\mathbf{X}) = A(\mathbf{X}^q) + \frac{1}{2}(\mathbf{X} - \mathbf{X}^q)_\alpha (\mathbf{X} - \mathbf{X}^q)_\beta \left.\frac{\partial^2 A(\mathbf{X})}{\partial \mathbf{X}_\alpha \, \partial \mathbf{X}_\beta}\right|_{\mathbf{X}^q}$$

$$+ \text{higher order terms}$$

$$= A(\mathbf{X}^q) + \frac{1}{2}(\mathbf{X} - \mathbf{X}^q)_\alpha (\mathbf{X} - \mathbf{X}^q)_\beta \, \mathbf{M}_{\alpha,\beta}$$

$$+ \text{higher order terms}. \tag{3.7}$$

To this order the expected value of $G(\mathbf{X})$ is

$$\langle G(\mathbf{X}) \rangle \approx \frac{\sum_{q=1}^{q_{max}} G(\mathbf{X}^q) \, c_q}{\sum_{q=1}^{q_{max}} c_q} \tag{3.8}$$

with

$$c_q = \exp -[A(\mathbf{X}^q)] \frac{1}{\sqrt{\det \mathbf{M}(\mathbf{X}^q))}}, \tag{3.9}$$

and

$$\mathbf{M}_{\alpha\,\beta}(\mathbf{X}^q) = \left.\frac{\partial^2 A(\mathbf{X})}{\partial X_\alpha \, \partial X_\beta}\right|_{\mathbf{X}^q}. \tag{3.10}$$

We order the action values along the paths so $A(\mathbf{X}^1) \leq A(\mathbf{X}^2) \ldots \leq A(\mathbf{X}^{q_{max}})$. The smaller action levels yield larger conditional probability values as $A(\mathbf{X}) = -\log[P(\mathbf{X}|\mathbf{Y})]$. If $A(\mathbf{X}^1)$ is much smaller than the other action levels, we see that it exponentially dominates the expected value:

$$\langle G(\mathbf{X}) \rangle \approx G(\mathbf{X}^1) + O(\exp[-A(\mathbf{X}^2) + A(\mathbf{X}^1)]), \tag{3.11}$$

so Laplace's method directs our attention to the paths with the smallest minimum actions.

Once we have found the path with the global minimum action, we can calculate corrections to Eq. (3.11) in two ways: (1) we may include, as a Taylor expansion around $A(\mathbf{X}^1)$, the cubic and higher order terms in the action (Ye et al. (2015b); Ye (2016)), and (2) we may include additional paths, $\mathbf{X}^2, \mathbf{X}^3, \ldots$.

This perturbation series is familiar in quantum field theory and statistical physics (Zinn-Justin (2002); Hochberg et al. (1999); Schwartz (2014)). It may, or may not, converge, but often much may be deduced about properties of the full perturbation series, despite the remarkable complexity and combinatorics of the individual terms, each of which is a Gaussian integral with powers of $(\mathbf{X} - \mathbf{X}^q)_\alpha$ depending on $A(\mathbf{X})$ and $G(\mathbf{X})$.

3.2.1 Laplace's Method Is NP-Complete

Another *caveat*, a troublesome one, is that the search for the global minimum of a nonlinear function of $\mathbf{X}$, such as our action $A(\mathbf{X})$, is NP-complete (Murty and Kabadi (1987); Garey and Johnson (1990)).

To discuss the implication of a problem's being "NP-Complete" we quote here a *slightly edited* commentary from the Encyclopedia Britannica (Gupta (2008)) entry on the subject from 2008:

> **NP-complete problem**, *any of a class of computational problems for which no efficient solution algorithm has been found. Many significant computer-science problems belong to this class.*
>
> *Easy problems can be solved by algorithms that run in polynomial time; i.e., for a problem of size N, the time or number of steps needed to find the solution is a polynomial function of N. Algorithms for solving hard, or intractable, problems, require times that are exponential in N. Polynomial-time algorithms are considered to be efficient, while exponential-time algorithms are considered inefficient.*
>
> *A problem is called NP (nondeterministic polynomial) if its solution can be guessed and verified in polynomial time; nondeterministic means that no particular rule is followed to make the guess. If a problem is NP and all other NP problems are polynomial-time reducible to it, the problem is NP-complete. Finding an efficient algorithm for any NP-complete problem implies that an efficient algorithm can be found for all such problems, since any problem belonging to this class can be recast into any other member of the class. It is not known whether any polynomial-time algorithms will ever be found for NP-complete problems, and determining whether these problems are tractable or intractable remains one of the most important questions in theoretical computer science.*

The implication for the problem at hand, finding the global minimum of a nonlinear function $A(\mathbf{X})$, (here we at least have $A(\mathbf{X}) \geq 0$), we interpret the absence of a general polynomial time search algorithm to mean that absent a special aspect of the nonlinear function whose global minimum we seek, we cannot effectively solve the general problem as the dimension of $\mathbf{X}$ becomes too large, suggesting, but by no means proving, that in some special cases there may be a problem class specific algorithm to do the task.

We will suggest such an algorithm in a later chapter. We will present it, prove nothing, but make the bold statement it does the job at hand, and then we suggest, by some examples, that it appears to do the job. We welcome a proof, but we'll ask again later. At this time (2020) we have no proof.

Just as a note here, when one mentions that searching for the minimum of a nonlinear function is NP-complete to friendly applied mathematicians, they furl their brows with concern. The same phrase mentioned to equally friendly Physicists gets a 'don't worry, mate' you are probably close enough. Take your choice.

3.3 The Euler-Lagrange Equations for the Standard Model: Continuous Time

We must approximate the expected value integrals numerically using discrete time methods where $t \rightarrow t_n = t_0 + n\Delta t$. Many aspects of the problem are revealed, however, in a continuous time framework. So we will start there.

We now take the limit $\Delta t \rightarrow 0$ to turn nonlinear discrete time maps $\mathbf{x}(n) \rightarrow \mathbf{x}(n+1)$ into differential equations for $\mathbf{x}(t)$. When numerical evaluation is called for, we go back to discrete time.

There is some subtlety in taking this limit, but it is well described in Durr and Bach (1978); Jouvet and Phythian (1979); Hochberg et al. (1999); Zinn-Justin (2002). The net result is an action that reads

$$A(\mathbf{X}) = \int_{t_0}^{t_{final}} dt\, L\left(\mathbf{x}(t), \frac{d\mathbf{x}(t)}{dt}, t\right)$$

in which

$$L\left(\mathbf{x}(t), \frac{d\mathbf{x}(t)}{dt}, t\right) = \left\{ \frac{R_m(t)}{2} \sum_{l=1}^{L} (y_l(t) - x_l(t))^2 + \right.$$

$$+ \sum_{a=1}^{D} \frac{R_f(a)}{2} \left(\frac{dx_a(t)}{dt} - F_a(\mathbf{x}(t), \boldsymbol{\theta}, t) \right)^2$$

$$\left. + \frac{1}{2} \nabla \cdot \mathbf{F}(\mathbf{x}(t), \boldsymbol{\theta}, t) \right\} - \log[P(\mathbf{x}(0)]$$

$$= \sum_{a=1}^{D} \frac{R_f(a)}{2} \left(\left(\frac{dx_a(t)}{dt} \right)^2 - 2 \frac{dx_a(t)}{dt} F_a(\mathbf{x}(t), \boldsymbol{\theta}, t) \right) + \chi(\mathbf{x}(t), t),$$

where

$$\chi(\mathbf{x}(t), t) = \frac{R_m(t)}{2} \sum_{l=1}^{L} (y_l(t) - x_l(t))^2 + \frac{1}{2} \nabla \cdot \mathbf{F}(\mathbf{x}(t), \boldsymbol{\theta}, t) - \log[P(\mathbf{x}(0)]$$

$$+ \sum_{b=1}^{D} \frac{R_f(b)}{2} \mathbf{F}_b(\mathbf{x}(t), t) \cdot \mathbf{F}_b(\mathbf{x}(t), \boldsymbol{\theta}, t). \tag{3.12}$$

From the work in Durr and Bach (1978); Jouvet and Phythian (1979); Hochberg et al. (1999); Zinn-Justin (2002), we see that the only change in the action as $\Delta t \to 0$ is the addition of the term $[\nabla \cdot \mathbf{F}(\mathbf{x}(t)]/2$.

The measurement error term involves $R_m(t)$, which, using the notation in Fig. (2.1), vanishes except at the observation times $t = \tau_k = n_k \Delta t$; $k = 1, 2, 3 \ldots, F$. We have also allowed for explicit time dependence in $\mathbf{F}(\mathbf{x}(t), \boldsymbol{\theta}, t)$. In SDA this is in the observations $\mathbf{y}(t)$.

The condition for a minimum of this action is the requirement that $\delta A(\mathbf{X}) = 0$, namely that the path $\mathbf{X}$ is an extremum of the action. This requirement, along with a partial integration in time, leads us to (Gelfand and Fomin (1963); Kirk (1970); Liberzon (2012); Kot (2014)) the Euler-Lagrange equations

$$\frac{d}{dt} \left\{ \frac{\partial L(\mathbf{x}(t), d\mathbf{x}(t)/dt, t)}{\partial \dot{x}_a(t)} \right\} = \frac{\partial L(\mathbf{x}(t), d\mathbf{x}(t)/dt, t)}{\partial x_a(t)}, \tag{3.13}$$

in which $\dot{\mathbf{x}}(t) = d\mathbf{x}(t)/dt$, and the accompanying boundary conditions at t_0 and t_{final} also must be satisfied to achieve $\delta A(\mathbf{X}) = 0$:

$$\delta \mathbf{x}(t) \cdot \left. \frac{\partial L(\mathbf{x}(t), d\mathbf{x}(t)/dt, t)}{\partial \dot{\mathbf{x}}(t)} \right|_{t_0}^{t_{final}} = 0$$

$$\delta \mathbf{x}(t) \cdot \mathbf{p}(t) \Big|_{t_0}^{t_{final}} = 0. \tag{3.14}$$

We use the *canonical momentum* (Goldstein et al. (2002); Arnol'd (1989)) in later considerations, and we introduce it now:

$$p_a(t) = \frac{\partial L(\mathbf{x}(t), d\mathbf{x}(t)/dt), t)}{\partial \dot{x}_a(t)}. \tag{3.15}$$

For the standard model this is

$$p_a(t) = R_f(a) \left[\frac{dx_a(t)}{dt} - F_a(\mathbf{x}(t), \boldsymbol{\theta}, t) \right]. \tag{3.16}$$

Eq. (3.14) is used in classical mechanics as an initial value problem where $\mathbf{x}(t_0)$ and $\dot{\mathbf{x}}(t_0)$ are given, and the dynamical equations are integrated forward in time.

In evaluating expected values Eq. (3.1), we integrate over all allowed values of the components of the path vector $\mathbf{X}$ including $\mathbf{x}(t_0)$ and $\mathbf{x}(t_{final})$, so $\delta \mathbf{x}(t_0) \neq 0$

and $\delta\mathbf{x}(t_{final}) \neq 0$, and we conclude that the appropriate boundary conditions for evaluating expected values are as follows (Gelfand and Fomin (1963); Kirk (1970); Liberzon (2012); Kot (2014)):

$$p_a(t_0) = p_a(t_{final}) = 0; \quad a = 1, 2, \ldots, D. \tag{3.17}$$

This is a two point boundary value problem (Greengard and Rokhlin (1991); Press et al. (2007); Gelfand and Fomin (1963)), which has a solid basis of working algorithms to use in solving them.

3.3.1 Lagrangian Coordinates $\{\mathbf{x}(t), \dot{\mathbf{x}}(t)\}$ and the EL Equation

For the standard model, the Euler-Lagrange equations take the form

$$R_f\left[\frac{d}{dt}\delta_{ab} + DF_{ab}(\mathbf{x}(t))\right]\left[\frac{dx_b(t)}{dt} - F_b(\mathbf{x}(t), \boldsymbol{\theta}, t)\right] = \frac{\partial \chi(\mathbf{x}(t) - \mathbf{y}(t))}{\partial x_a(t)}$$

$$\frac{d^2 x_a(t)}{dt^2} - \Omega_{ab}(\mathbf{x}(t))\dot{x}_b(t) = \frac{\partial\left[\frac{\chi(\mathbf{X}(t) - \mathbf{y}(t))}{R_f} + \frac{\mathbf{F}(\mathbf{X}(t))^2}{2}\right]}{\partial x_a(t)}$$

$$+ \frac{\partial F_a(\mathbf{x}(t), t)}{\partial t} \tag{3.18}$$

where we have

$$DF_{ab}(\mathbf{x}) = \partial F_a(\mathbf{x}, \boldsymbol{\theta}, t)/\partial x_b; \quad \Omega_{ab} = DF_{ab}(\mathbf{x}(t), \boldsymbol{\theta}, t) - DF_{ba}(\mathbf{x}(t));$$

$$\chi(\mathbf{x}(t) - \mathbf{y}(t)) = \sum_{r=1}^{L} \frac{R_m(r, t)}{2}\left(x_r(t) - y_r(t))^2\right)$$

$$+ \frac{1}{2}\nabla \cdot \mathbf{F}(\mathbf{x}(t), \boldsymbol{\theta}, t) - \log[P[\mathbf{x}(0)]. \tag{3.19}$$

The Euler-Lagrange equation for the Standard Model resembles the motion of a charged object in a D-dimensional magnetic field and an electric field in D-dimensions (Abarbanel et al. (2018)). The term analogous to the magnetic field in the EL equation is the skew symmetric matrix term in the EL equation. We see that it is perpendicular to $v_a(t)$; $v_a(t) = \frac{dx_a(t)}{dt}$.

$$\Omega_{ab}(\mathbf{x}(t))v_b(t) = \left\{\frac{\partial F_a(\mathbf{x}(t), t)}{\partial x_a(t))} - \frac{\partial F_b(\mathbf{x}(t), t)}{\partial x_a(t)}\right\}v_b(t). \tag{3.20}$$

The analog electric field is

$$\mathcal{E}_b(\mathbf{x}, t) = \frac{1}{R_f(b)}\frac{\partial \chi(\mathbf{x}(t), t)}{\partial x_b(t)} + \frac{\partial F_b(\mathbf{x}(t), \boldsymbol{\theta}, t)}{\partial t}. \tag{3.21}$$

The analog is that the vector potential is $\mathbf{F}(\mathbf{x}(t), \boldsymbol{\theta}, t)$ and $\chi(\mathbf{x}(t), t)$ is the scalar potential.

If we make the 'gauge' transformation $F_a(\mathbf{x}, \boldsymbol{\theta}, t) \rightarrow F_a(\mathbf{x}, \boldsymbol{\theta}, t) + \nabla_a \psi(\mathbf{x}, t)$, then Ω_{ab} is unchanged. The 'electric field' becomes

$$
\mathcal{E}_b(\mathbf{x}, t) \rightarrow \mathcal{E}_b(\mathbf{x}, t) + \nabla_b \left[\mathbf{F}(\mathbf{x}, t) \cdot \nabla \psi(\mathbf{x}, t) + \frac{\nabla \psi(\mathbf{x}, t))^2}{2} \right.
$$
$$
\left. + \frac{\nabla^2 \psi(\mathbf{x}(t), t)}{R_f} + \frac{\partial \psi(\mathbf{x}, t)}{\partial t} \right], \qquad (3.22)
$$

and this 'electric field' and the equations of motion are unchanged if $\psi(\mathbf{x}, t)$ satisfies

$$
\frac{\partial \psi(\mathbf{x}, t)}{\partial t} + \mathbf{F}(\mathbf{x}, \boldsymbol{\theta}, t) \cdot \nabla \psi(\mathbf{x}, t) + \frac{(\nabla \psi(\mathbf{x}, t))^2}{2}
$$
$$
+ \frac{\nabla^2 \psi(\mathbf{x}(t), t)}{R_f} = g_{time}(t), \qquad (3.23)
$$

and $g_{time}(t)$ is any function of time alone.

Along with an invariance such as this, there is a conserved current. In this case it happens to be local in $\mathbf{x}$ space and is the conservation of number of particles. Were we discussing recurrent networks, more structure would be involved.

3.3.2 Hamiltonian Canonical Coordinates and the EL Equation

In $(\mathbf{x}, \mathbf{p})$ space the Euler-Lagrange (E-L) equation for the Standard Model takes the form, using Eq. (3.16):

$$
\frac{dx_a(t)}{dt} = F_a(\mathbf{x}(t)) + \frac{p_a(t)}{R_f}
$$
$$
\frac{dp_a(t)}{dt} = -\frac{\partial F_b(\mathbf{x}(t))}{\partial x_a(t)} p_b(t) + \frac{\partial \left\{ \frac{R_m(t)}{2} \sum_{l=1}^{L} (y_l(t) - x_l(t))^2 + \frac{1}{2} \nabla \cdot \mathbf{F}(\mathbf{x}(t)) \right\}}{\partial x_a(t)}
$$
$$
- \frac{\partial \log[P(\mathbf{x}(0))]}{\partial x_a(0)}. \qquad (3.24)
$$

We now make a very useful change of coordinates, $\{\mathbf{x}(t), \dot{\mathbf{x}}(t)\} \rightarrow \{\mathbf{x}(t), \mathbf{p}(t)\}$. Using $\{p_a(t) = \partial L(\mathbf{x}(t), \dot{\mathbf{x}}(t), t)/\partial \dot{x}_a(t)\}$ as just defined. The $\{\mathbf{x}(t), \mathbf{p}(t)\}$ are called canonical coordinates, and we use them to introduce the Hamiltonian $H(\mathbf{x}(t), \mathbf{p}(t), t)$:

$$
H(\mathbf{x}(t), \mathbf{p}(t), t) = \dot{\mathbf{x}}(t) \cdot \mathbf{p}(t) - L(\mathbf{x}(t), \dot{\mathbf{x}}(t), t).
$$

For the standard model, this leads us to

$$
H(\mathbf{x}(t), \mathbf{p}(t), t) = \frac{\mathbf{p}(t) \cdot \mathbf{p}(t)}{2R_f} + \mathbf{p}(t) \cdot \mathbf{F}(\mathbf{x}(t), \boldsymbol{\theta}, t) - \chi(\mathbf{x}(t), t),
$$

where

$$\chi(\mathbf{x}(t), \mathbf{y}(t)) = \frac{R_m(t)}{2} \sum_{l=1}^{L} (y_l(t) - x_l(t))^2$$

$$+ \frac{1}{2} \nabla \cdot \mathbf{F}(\mathbf{x}(t), \boldsymbol{\theta}, t) - \log[P(\mathbf{x}(0)] + \frac{1}{2} \mathbf{F}(\mathbf{x}(t), \boldsymbol{\theta}, t) \cdot \mathbf{F}(\mathbf{x}(t), \boldsymbol{\theta}, t). \quad (3.25)$$

Hamilton's equations of motion are found to be (Arnol'd (1989); Goldstein et al. (2002))

$$\frac{dx_a(t)}{dt} = \frac{\partial H(\mathbf{x}(t), \mathbf{p}(t), t)}{\partial \mathbf{p}_a(t)} = \frac{p_a(t)}{R_f} + F_a(\mathbf{x}(t), \boldsymbol{\theta}, t)$$

$$\frac{dp_a(t)}{dt} = -\frac{\partial H(\mathbf{x}(t), \mathbf{p}(t), t)}{\partial \mathbf{x}_a(t)}$$

$$= -p_b(t) \frac{\partial F_b(\mathbf{x}(t), \boldsymbol{\theta}, t)}{\partial x_a(t)} + \frac{\partial \chi(\mathbf{x}(t), t)}{\partial x_a(t)}. \quad (3.26)$$

While this transformation to canonical coordinates appears to have some scent of mystery to it, the reward is a set of coordinates in which the Hamiltonian is conserved under evolution in time, if it is itself time independent, and it is often the energy of the dynamical system it represents.

The motion in $\{\mathbf{x}(t), \mathbf{p}(t)\}$ space also exposes in a clear and useful manner the underlying symplectic behavior (Arnol'd (1989); Goldstein et al. (2002)) of those orbits. We will further investigate this aspect of variational principles in the next section and again in Chapter 6.

If we take the deterministic limit $R_f \to \infty$ with $\mathbf{p}(t)$ held fixed, Hamilton's equations become

$$\frac{dx_b(t)}{dt} = F_b(\mathbf{x}(t), \boldsymbol{\theta}, t)$$

and

$$\frac{dp_b(t)}{dt} = -\frac{\partial F_a(\mathbf{x}(t), \boldsymbol{\theta})}{\partial x_b(t)} p_a(t) + \frac{\partial \chi(\mathbf{x}(t) - \mathbf{y}(t))}{\partial x_b(t)}. \quad (3.27)$$

These equations are familiar in two contexts: (1) they are the inhomogeneous equations ($\chi(\mathbf{x} - \mathbf{y}) = 0$) of *optimal control* theory as developed by Pontryagin (1959) and Gelfand and Fomin (1963), and (2) they are the equations of machine learning where time $\to$ layer (Goodfellow et al. (2016); Abarbanel et al. (2018)). While these, as in the Lagrangian formulation (Gelfand and Fomin (1963)), are two point boundary value problems, the solution methods used by investigators in different research areas can vary significantly. In machine learning one starts at t_{final} and propagates the solution back to t_0. The explicit presence of the Jacobian $\frac{\partial F_a(\mathbf{X}(t), \boldsymbol{\theta})}{\partial x_b(t)}$ in the equation for the canonical momentum can result in sensitivities to errors, as in other nonlinear dynamical systems, or even chaotic solutions to the equations. We will revisit these equations in Chapter 8.

3.3.3 Symplectic Symmetry of Variational Principles

We will use the discussion in this section in later chapters, but it is introduced here because of the context and to avoid "punitive pedagogy," in which all ideas are mentioned once and **only** once, regardless of their importance as much as possible.

In Hamiltonian mechanics, the state space, or phase space, is characterized (Goldstein et al. (2002)) by the 2D state variables $\{q_a(t), p_a(t)\}$; $a = 1, 2, \ldots, D$, the coordinates, and the canonical momentum. The Hamiltonian function $H(\mathbf{q}(t), \mathbf{p}(t))$ generates motion in this phase space via

$$\frac{dq_a(t)}{dt} = \frac{\partial H(\mathbf{q}(t), \mathbf{p}(t))}{\partial p_a(t)} \quad \frac{dp_a(t)}{dt} = -\frac{\partial H(\mathbf{q}(t), \mathbf{p}(t))}{\partial q_a(t)}. \tag{3.28}$$

The time evolution of an arbitrary function $A(\mathbf{q}(t), \mathbf{p}(t))$ on the phase space is given by

$$\frac{dA(\mathbf{q}(t), \mathbf{p}(t))}{dt} = \frac{\partial A(\mathbf{q}(t), \mathbf{p}(t))}{\partial q_a(t)} \frac{dq_a(t)}{dt} + \frac{\partial A(\mathbf{q}(t), \mathbf{p}(t))}{\partial p_a(t)} \frac{dp_a(t)}{dt}$$

and using Eq. (3.28) gives us

$$\frac{dA(\mathbf{q}(t), \mathbf{p}(t))}{dt} = \frac{\partial A(\mathbf{q}(t), \mathbf{p}(t))}{\partial q_a(t)} \frac{\partial H(\mathbf{q}(t), \mathbf{p}(t))}{\partial p_a(t)}$$

$$-\frac{\partial A(\mathbf{q}(t), \mathbf{p}(t))}{\partial p_a(t)} \frac{\partial H(\mathbf{q}(t), \mathbf{p}(t))}{\partial p_a(t)}$$

$$\frac{dA(\mathbf{q}(t), \mathbf{p}(t))}{dt} = \left\{ A(\mathbf{q}(t), \mathbf{p}(t)), H(\mathbf{q}(t), \mathbf{p}(t)) \right\} \bigg|_{\mathbf{q}(t), \mathbf{p}(t)}. \tag{3.29}$$

This introduces the **Poisson bracket** of two functions, $A(\mathbf{q}, \mathbf{p})$ and $B(\mathbf{q}, \mathbf{p})$, on phase space

$$\left\{ A(\mathbf{q}, \mathbf{p}), B(\mathbf{q}, \mathbf{p}) \right\} \bigg|_{\mathbf{q}, \mathbf{p}} = \frac{\partial A(\mathbf{q}, \mathbf{p})}{\partial q_a} \frac{\partial B(\mathbf{q}, \mathbf{p})}{\partial p_a} - \frac{\partial A(\mathbf{q}, \mathbf{p})}{\partial p_a} \frac{\partial B(\mathbf{q}, \mathbf{p})}{\partial p_a}. \tag{3.30}$$

The Poisson brackets of the phase space coordinates are

$$\{q_a, p_b\} \bigg|_{\mathbf{q}, \mathbf{p}} = -\{p_b, q_a\} \bigg|_{\mathbf{q}, \mathbf{p}} = \delta_{ab}; \ \{q_a, q_b\} \bigg|_{\mathbf{q}, \mathbf{p}} = \{p_a, p_b\} \bigg|_{\mathbf{q}, \mathbf{p}} = 0. \tag{3.31}$$

When we make a canonical transformation of coordinates $\{\mathbf{q}, \mathbf{p}\} \rightarrow \{\mathbf{Q}(\mathbf{q}, \mathbf{p}), \mathbf{P}(\mathbf{q}, \mathbf{p})\}$, for example by solving Hamilton's equations, the Poisson brackets of $\{\mathbf{Q}, \mathbf{P}\}$ are the same:

$$\{Q_a, P_b\} \bigg|_{\mathbf{Q}, \mathbf{P}} = -\{P_b, Q_a\} \bigg|_{\mathbf{Q}, \mathbf{P}} = \delta_{ab}; \ \{Q_a, Q_b\} \bigg|_{\mathbf{Q}, \mathbf{P}} = \{P_a, P_b\} \bigg|_{\mathbf{Q}, \mathbf{P}} = 0.$$

$$\tag{3.32}$$

Movement through phase space $\{\mathbf{q}, \mathbf{p}\}$ using the rules of Hamiltonian mechanics, labeled by time, is a transformation from an initial location $\{\mathbf{q}, \mathbf{p}\}$ to a different location $\{\mathbf{Q}(\mathbf{q}, \mathbf{p}), \mathbf{P}(\mathbf{q}, \mathbf{p})\}$. Movement further along $\{\mathbf{q}, \mathbf{p}\} \rightarrow \{\mathbf{Q}(\mathbf{q}, \mathbf{p}), \mathbf{P}(\mathbf{q}, \mathbf{p}))$ $\rightarrow \{\mathbf{Q}'(\mathbf{Q}, \mathbf{P}), \mathbf{P}'(\mathbf{Q}, \mathbf{P})\} \rightarrow \ldots$ following the Poisson bracket rules of Hamiltonian mechanics preserves the Poisson brackets as expressed by

$$
\left\{ A(\mathbf{q}, \mathbf{p}), B(\mathbf{q}, \mathbf{p}) \right\} \bigg|_{\mathbf{q}, \mathbf{p}} = \left\{ A(\mathbf{Q}, \mathbf{P}), B(\mathbf{Q}, \mathbf{P}) \right\} \bigg|_{\mathbf{Q}, \mathbf{P}}
$$

$$
= \left\{ A(\mathbf{Q}', \mathbf{P}'), B(\mathbf{Q}', \mathbf{P}') \right\} \bigg|_{\mathbf{Q}', \mathbf{P}'} = \ldots \quad (3.33)
$$

This is summarized by saying that each transformation via Hamilton's rules $\{\mathbf{q}, \mathbf{p}\} \rightarrow \{\mathbf{Q}(\mathbf{q}, \mathbf{p}), \mathbf{P}(\mathbf{q}, \mathbf{p})\} \rightarrow \{\mathbf{Q}'(\mathbf{Q}, \mathbf{P}), \mathbf{P}'(\mathbf{Q}, \mathbf{P})\} \rightarrow \ldots$ is *canonical* in that it goes from one location in phase space to another via a sequence of Hamiltonia: $H_1(\mathbf{q}, \mathbf{p}) = H_2(\mathbf{Q}(\mathbf{q}, \mathbf{p}), \mathbf{P}(\mathbf{q}, \mathbf{p})) \ldots$. At each step, further movement along the phase space trajectory is generated by the scalar Hamiltonian evaluated at the phase space location reached by the previous step. So, the motion is generated by Hamiltonian dynamics as one proceeds. This *defines* canonical transformations.

The movement through phase space labeled by time generated by Hamilton's equations $\{\mathbf{q}(t_0), \mathbf{p}(t_0)\} \rightarrow \{\mathbf{q}(t_1), \mathbf{p}(t_t)\} \rightarrow \ldots$ is a sequence of canonical transformations. Each such canonical transformation takes us to a new point in phase space, from which further motion in time is also generated by Hamilton's rules. Each such transformation preserves the Poisson brackets among functions on phase space.

With all that structure, one may not be surprised to learn there is an underlying symmetry associated with all this. It happens to be the symmetry of *symplectic* transformations. A symplectic transformation $\mathbf{S}$ satisfies

$$
\mathbf{J} = \mathbf{S}^T \mathbf{J} \mathbf{S}, \quad (3.34)
$$

in which

$$
\mathbf{J} = \begin{pmatrix} 0 & \mathcal{I} \\ -\mathcal{I} & 0 \end{pmatrix} \quad (3.35)
$$

and $\mathbf{S}^T$ is the transpose of the matrix $\mathbf{S}$.

In Eq. (3.35) $\mathcal{I}$ is a $D \times D$ unit matrix, and 0 is a $D \times D$ null matrix. $\mathbf{J}$ and $\mathbf{S}$ are $2D \times 2D$ matrices.

A canonical transformation $\{\mathbf{q}, \mathbf{p}\} \rightarrow \{\mathbf{Q}(\mathbf{q}, \mathbf{p}), \mathbf{P}(\mathbf{q}, \mathbf{p})\}$ enables us to form the $2D \times 2D$ Jacobian matrix

$$
\mathbf{S} = \begin{pmatrix} \frac{\partial \mathbf{Q}(\mathbf{q}, \mathbf{p})}{\partial \mathbf{q}} & \frac{\partial \mathbf{Q}(\mathbf{q}, \mathbf{p})}{\partial \mathbf{p}} \\ \frac{\partial \mathbf{P}(\mathbf{q}, \mathbf{p})}{\partial \mathbf{q}} & \frac{\partial \mathbf{P}(\mathbf{q}, \mathbf{p})}{\partial \mathbf{p}} \end{pmatrix}. \quad (3.36)
$$

With only a little extra suffering, we can form

$$\mathbf{S}^T \mathbf{J} \mathbf{S} \tag{3.37}$$

and establish it is

$$\begin{pmatrix} 0 & \{\mathbf{Q(q, p)}, \mathbf{P(q, p)}\}\big|_{\mathbf{q, p}} \\ -\{\mathbf{Q(q, p)}, \mathbf{P(q, p)}\}\big|_{\mathbf{q, p}} & 0 \end{pmatrix} \tag{3.38}$$

which is $\mathbf{J}$ as $\{Q_a, P_b\} = \delta_{ab}$.

So we can finally conclude, after a deep breath, that since moving through phase space following Hamilton's rules is a sequence of canonical transformations, it is also a sequence of steps that satisfy a symplectic symmetry. Each of these steps preserves the Poisson brackets at each stage.

Now, the reason we have done all of this is to emphasize that, while this is *automatically* true for continuous time evolution of Hamiltonian dynamics, it is *not* automatically so for discrete time Hamiltonian dynamics where time evolution proceeds in finite time steps $\Delta t \neq 0$. This alerts us that when we come to numerical evaluations of expectation values or other interesting quantities in data assimilation, we must be more careful in order to preserve the symplectic symmetry of the problem at hand.

We will have to impose symplecticity on moving Hamilton's equations along in discrete time to enjoy the benefits of the symplectic symmetry. Two such benefits are these: (a) conservation of the Hamiltonian along the orbits in phase space and (b) conservation of phase space volume along those orbits – as noted: both are automatically respected in continuous time, but, alas, not automatically in discrete time.

4

Using Waveform Information

4.1 Inconsistency in the Standard Model Action

The formulation of the Standard Model has an inconsistency, mentioned earlier, and we address this matter now.

The general development of the action in Eq. (2.6) allows the information transfer term,

$$\sum_{n=1}^{N} \log[P(\mathbf{y}(n)|\mathbf{x}(n), \mathbf{X}(n-1), \mathbf{Y}(n-1))], \tag{4.1}$$

to depend on all previous states and all previous measurements. When we specialized to the standard model action

$$A(\mathbf{X}) = \sum_{n=1}^{N} \frac{R_m(n)}{2} \sum_{l=1}^{L} (x_l(n) - y_l(n))^2 + \text{model error term}, \tag{4.2}$$

we neglected the dependence on earlier measurements in Eq. (4.1) and approximated $P(\mathbf{y}(n)|\mathbf{x}(n), \mathbf{X}(n-1), \mathbf{Y}(n-1))$ as $P(\mathbf{y}(n)|\mathbf{x}(n))$. Making this approximation is effectively saying that at each measurement time τ_k, the $y_l(\tau_k)$ are *independent* of all other measurements at other observation times $\tau_{j \neq k}$.

However, in the measurement error term we ask that the values of the model state variables, $x_l(\tau_k)$, be as close as we can make them to the observations, $y_l(\tau_k)$, namely, $|x_l(\tau_k) - y_l(\tau_k)| \approx 1/\sqrt{R_m}$, which is to say they are the same, within the measurement errors.

The $x_l(t)$ are dynamically connected to each other via the model equations of motion $\mathbf{x}(n+1) = \mathbf{f}(\mathbf{x}(n), \boldsymbol{\theta})$, whether they are observed or unobserved. That would require the $y_l(t)$ also to be dynamically connected to each other, and in this approximation, they are not.

4.2 Time Delay State Vectors and Data

In the analysis of nonlinear dynamical time series another face of this question was addressed some years ago (Takens (1981); Eckmann and Ruelle (1985); Abarbanel (1996); Kantz and Schreiber (2004)). In the common situation where one has a single observed scalar time series $s(t_n)$, the method of *time delay embedding* is used to create D_E-dimensional time delay vectors $\{s(t_n), s(t_n - \Delta), \ldots, s(t_n - (k - 1)\Delta), \ldots, s(t_n - (D_E - 1)\Delta))\}$ that add information about the dynamical connections among observations which evolve into $s(t_n)$, and they do so using all of the state variables of the source $s(t)$ present in the function $f(s(t), \theta) = ds(t)/dt$:

$$s(t_n - \Delta) \approx s(t_n) + \int_{t_n}^{t_n - \Delta} dt' f(s(t'), \theta)$$
$$\approx s(t_n) - \Delta f(s(t_n), \theta), \tag{4.3}$$

for small Δ.

We can see explicitly how knowledge of $s(t)$, involving all components of the state of the observed system, at earlier times than t_n, is introduced into $s(t_n)$. This is how the use of time delay coordinates adds information from unobserved state variables in the full state $s(t)$ and parameters θ to the progression in time of the observed scalar $s(t)$.

Determining the time lag Δ and the dimension of the proxy space D_E is well explored and well documented (Takens (1981); Aeyels (1981a,b); Eckmann and Ruelle (1985); Abarbanel (1996); Kantz and Schreiber (2004)).

The connections among the model state variables $x(t)$ are given by the underlying differential equations in time or the layer-to-layer rules in machine learning (Abarbanel et al. (2018)).

The idea here is not to necessarily accurately approximate the derivatives $dy(t)/dt$ and higher derivatives in time. Through these connections among the observations one can gain enough information to achieve the connections of the model states $x(t)$ to the measurements.

As in the situation with phase space reconstruction utilizing time delay embedding (Abarbanel (1996); Kantz and Schreiber (2004)), the goal is to add other, not quite independent, coordinates to the characterization we have of the evolving time dependence of the observations to be associated with the correlated, evolving time dependence of the state of the model. The additional coordinates present more information to the model and as such permit more accurate estimations of the model parameters θ and the unmeasured state variables. The information cost in the estimation of a parameter or a state variable is analyzed by Rissanen (1989).

In the differential equation version of the model, we gain the new information beyond knowledge of $\mathbf{x}(t)$ through the derivative $d\mathbf{x}(t)/dt$ which is equal to the vector field $\mathbf{F}(\mathbf{x}(t), \boldsymbol{\theta})$. We proceed with this guidance by initially approximating the first derivative of the measurements "$d\mathbf{y}(t)/dt$" at $t = \tau_i$ as

$$\frac{d\mathbf{y}(\tau_i)}{dt} \approx \frac{\mathbf{y}(\tau_i) - \mathbf{y}(\tau_{i-1})}{(n_i - n_{i-1})\Delta t}, \tag{4.4}$$

and we would like to take $\Delta t \to 0$.

The reality of collecting data at discrete times stands in the way. We cannot take $\Delta t \to 0$. We note, however, that the source of new information in this expression, Eq. (4.4), is $\mathbf{y}(\tau_{i-1})$, as we already know the other quantities in the approximation.

In data assimilation we observe L-dimensional quantities at each measurement time, $y_l(\tau_k)$; $l = 1, 2, \ldots, L$. Connecting up the observed information at τ_i with knowledge in the data at τ_i; $i \neq k$ is achieved by creating $L \cdot D_E$-dimensional state vectors, $S_{lk}(n) = x(t_n - (k-1)\rho) = x_l(n - (k-1)n_\rho)$; $k = 1, 2, \ldots D_E$; $l = 1, 2, \ldots, L$, composed of $x_l(t)$ and D_E of its time delays. The time delay ρ is also taken as an integer multiple of Δt: $\rho = n_\rho \Delta t$.

The notation we have used throughout this monograph is to write $\mathbf{x}(t_n) = \mathbf{x}(t_0 + n\Delta t) = \mathbf{x}(n)$. Formally, I suppose, we should have a different symbol for the state as a function of time to distinguish it from the state as a function of the dimensionless quantity 'n,' an integer, but, hoping it will not trouble the reader, we do not.

We write the components of the time delay state vectors as

$$S_{l1}(n) = x_l(n) = x_l(t_n)$$
$$S_{l2}(n) = x_l(n - n_\rho) = x_l(t_n - \rho) = x_l(t_0 + n\Delta t - n_\rho\Delta t)$$
$$S_{l3}(n) = x_l(n - 2n_\rho) = x_l(t_n - 2\rho)$$
$$\vdots$$
$$S_{lk}(n) = x_l(n - (k-1)n_\rho) = x_l(t_n - (k-1)\rho)$$
$$\vdots$$
$$S_{lD_E}(n) = x_l(n - (D_E - 1)n_\rho) = x_l(t_n - (D_E - 1)\rho), \tag{4.5}$$

in which $l = 1, 2, \ldots, L$.

We call these $S_{lk}(n)$ our *proxy state* variables, as their definition adds to the previous L-dimensional $x_l(n) = x_l(t_n)$ a set of specific time delayed state vectors $x_l(t_n - (k-1)\rho)$; $k = 2, 3, \ldots, D_E$ at time delays that are integer multiples of Δt, $\rho = n_\rho \Delta t$. This is the same time step used in the collection of the data and in the solution of the model equations. Using integer multiples of Δt is not required

(Takens (1981); Aeyels (1981a,b)), but this makes it possible to use most simply what we have already collected without imposing any additional interpolation or extrapolation.

We know the dynamical equation for the state vectors in the path $\mathbf{X}$: $x_l(t_{n+1}) = x_l(n+1) = f_l(\mathbf{x}(n), \boldsymbol{\theta})$, and this allows us to find equations of motion in discrete time for the components of the proxy state vectors $S_{lk}(n) = x_l(n-(k-1)\rho) = x_l(t_n - (k-1)\rho)$:

$$S_{l1}(n+n_\rho) = x_l(n+n_\rho) = [\mathbf{f}^{n_\rho}(\mathbf{x}(n))]_l$$
$$S_{l2}(n+n_\rho) = x_l(n) = S_{l1}(n)$$
$$S_{l3}(n+n_\rho) = x_l(n-1) = S_{l2}(n)$$
$$\vdots$$
$$S_{lk}(n+n_\rho) = x_l(n-(k-2)n_\rho) = S_{l(k-1)}(n)$$
$$\vdots$$
$$S_{lD_E}(n+n_\rho) = x_l(n-(D_E-2)n_\rho) = S_{l(D_E-1)}(n). \tag{4.6}$$

In these equations $[\mathbf{f}^{n_\rho}(\mathbf{x}(n))]_l$ is the lth component of $\mathbf{f}$ iterated n_ρ times: $\mathbf{f}^{n_\rho}(\mathbf{x}(n))$.

We also need to augment the measured quantities $y_l(\tau_r)$; $l = 1, 2, \ldots, L$; $r = 1, 2, \ldots, F$ using time delays. To this end we introduce the data time delay vector $Y_{lk}(r)$ with components

$$Y_{l1}(r) = y_l(\tau_r) = y_l(n_r),$$
$$Y_{l2}(r) = y_l(\tau_r - \rho) = y_l(n_r - n_\rho)$$
$$\vdots$$
$$Y_{lk}(r) = y_l(\tau_r - (k-1)\rho) = y_l(n_r - (k-1)n_\rho)$$
$$\vdots$$
$$Y_{lD_E}(n) = y_l(\tau_r - (D_E-1)\rho) = y_l(n_r - (D_E-1)n_\rho). \tag{4.7}$$

We ask, in the measurement error term of the action, that $S_{lk}(r) \approx Y_{lk}(r)$ for each component $k = 1, 2, \ldots, D_E$ and $l = 1, 2, \ldots, L$ at time τ_r.

The time delay or waveform augmented version of the standard model action in discrete time now takes the form

$$A(\mathbf{X}) = \sum_{r=1}^{F} \sum_{l=1}^{L} \sum_{k=1}^{D_E} \frac{R_m(r)}{2} \left\{ Y_{lk}(r) - S_{lk}(r) \right\}^2$$
$$+ \sum_{n=0}^{N-1} \sum_{a=1}^{D} \frac{R_f^{(x_a)}}{2} \left[x_a(n+1) - f_a(\mathbf{x}(n), \boldsymbol{\theta}) \right]^2. \tag{4.8}$$

This can be written as

$$A(\mathbf{X}) = \sum_{n=1}^{N}\sum_{l=1}^{L}\sum_{k=1}^{D_E} \frac{R_m(n)}{2} \left\{ Y_{lk}(n) - S_{lk}(n) \right\}^2$$

$$+ \sum_{n=0}^{N-1}\sum_{a=1}^{D} \frac{R_f^{(x_a)}}{2} \left[x_a(n+1) - f_a(\mathbf{x}(n), \boldsymbol{\theta}) \right]^2, \tag{4.9}$$

with the understanding that the precision $R_m(n) = 0$, except at the measurement times $t_n = \tau_k$.

4.3 "Old Nudging" for Proxy Vectors

If we let time become continuous, we may write an expression for $dS_{lk}(\mathbf{x}(t))/dt$ (Rey et al. (2014b); Rey (2017)):

$$\frac{dS_{lk}(\mathbf{x}(t))}{dt} = \sum_{a=1}^{D} \frac{\partial S_{lk}(\mathbf{x}(t))}{\partial x_a(t)} \frac{dx_a(t)}{dt}$$

$$+ \sum_{k'=1}^{D_E}\sum_{l'=1}^{L} g'_{lk,l'k'}(t) \left\{ Y_{l'k'}(t) - S_{l'k'}(t) \right\}$$

$$= \sum_{a=1}^{D} \frac{\partial S_{lk}(\mathbf{x}(t))}{\partial x_a(t)} F_a(\mathbf{x}(t), \boldsymbol{\theta}) + \sum_{k'=1}^{D_E}\sum_{l'=1}^{L} g'_{lk,l'k'}(t) \left\{ Y_{l'k'}(t) - S_{l'k'}(t) \right\}$$

or

$$\frac{dx_a(t)}{dt} = F_a(\mathbf{x}(t), \boldsymbol{\theta})$$

$$+ \sum_{k=1}^{D_E}\sum_{l=1}^{L}\sum_{k'=1}^{D_E}\sum_{l'=1}^{L} \frac{\partial x_a(t)}{\partial S_{lk}(\mathbf{x}(t))} g'_{lk,l'k'}(t) \left\{ Y_{l'k'}(t) - S_{l'k'}(t) \right\}$$

$$\frac{dx_a(t)}{dt} = F_a(\mathbf{x}(t), \boldsymbol{\theta}) + \sum_{a'=1}^{D} g_{aa'}(t)\delta x_{a'}(t), \tag{4.10}$$

where

$$\delta\mathbf{x}(t) = \frac{\partial \mathbf{x}}{\partial \mathbf{S}} g'(\mathbf{Y} - \mathbf{S}), \tag{4.11}$$

in matrix notation, and we added a scale factor $g_{aa'}(t)$ into the nudging term.

The interpretation of the nudging (or control) of the original dynamical equations $\dot{\mathbf{x}}(t) = \mathbf{F}(\mathbf{x}(t), \boldsymbol{\theta})$ is that the inhomogeneous term (right side) in Eq. (4.10) drives the proxy vector $S_{lk}(\mathbf{x}(t)) \to Y_{lk}(t)$ over time in an interval $[t_0, t_{final}]$. As it does that, it should be minimizing the cost function

$$\sum_{l=1}^{L}\sum_{k=1}^{D_E}(S_{kl}(\mathbf{x}(t)) - Y_{kl}(t))^2 + \sum_{a=1}^{D}\sum_{a'=1}^{D}g_{aa'}(t)g_{a'a}(t) \tag{4.12}$$

to zero. This synchronizes the proxy vectors $S_{kl}(\mathbf{x}(t))$ and $Y_{kl}(t)$ while sending the coupling or control to zero as well.

There is an interesting aspect to this formulation in proxy space in that the parameters $\boldsymbol{\theta}$ taken to satisfy the equation $d\boldsymbol{\theta}/dt = 0$ may be added to the list of dynamical equations and, through the term on the right hand side of Eq. (4.10), will now evolve in time and, as we will see, leads to the determination of the $\boldsymbol{\theta}$ as well as the unmeasured state variables.

4.4 L = 1; x_1(t) Is Observed

The power of adding waveform information is already visible, in the case $L = 1$ where we can observe only a single state variable, $x_1(t)$. We continue using information from the scalar time series of measurements $y(t_n)$ and its time delays as contained in Y_{1k}.

We establish synchronization on the synchronization manifold by comparing the D_E-dimensional data time delay vector $y_k(t) = \{y(t + (k - 1)\rho)\}$; $k = 1, 2, \ldots, D_E$ with the D_E-dimensional model output and its D_E time delays $x_1(t + (k-1)\rho)$; $k = 1, 2, \ldots, D_E$ through monitoring the *synchronization error*.

$$SE_s(t) = \frac{1}{D_E}\sum_{k=1}^{D_E}[Y_{1k}(t) - S_{1k}(\mathbf{x}(t))]^2,$$

or

$$SE_s(t) = \frac{1}{D_E}\sum_{k=1}^{D_E}[y(t + (k - 1)\rho) - x_1(t + (k - 1)\rho)]^2. \tag{4.13}$$

We expect that whatever the initial condition, $x_1(0)$, if D_E is large enough, the proxy vector $S_{1k}(\mathbf{x}(0)$ will converge to $Y_{1k}(t)$ as time goes by. This means we expect $SE_s(t) \to 0$.

The immediate challenge is interpreting the inverse $\frac{\partial x_a(t)}{\partial S_{1k}(\mathbf{X}(t))}$ of the $D_E \times D$ rectangular matrix $\frac{\partial S_{1k}(\mathbf{X}(t))}{\partial x_a(t)}$.

To implement this control (nudging) approach, we integrate the $g_{aa'} = 0$ equations of motion

$$\frac{d\bar{x}_a(t)}{dt} = F_a(\bar{\mathbf{x}}(t), \boldsymbol{\theta}); \quad a = 1, 2, \ldots, D, \tag{4.14}$$

from time $t - (D_E - 1)\rho$ to time t to construct the vector $S_{1k}(\bar{\mathbf{x}}(t))$. We introduce the notation $\bar{\mathbf{x}}(t)$ to identify the solution with $g_{aa'} = 0$; we also call this the *uncoupled* dynamics.

To evaluate the Jacobian $\partial\mathbf{x}/\partial\mathbf{S}$ appearing in Eq. (4.10), at each time step t, integrate the variational equation (Teschl (2010); Chapter 2),

$$\frac{d}{dt}\Phi_{ab}(t) = \sum_{c=1}^{D}\left.\frac{\partial F_a(\mathbf{x}(t))}{\partial x_c(t)}\right|_{\bar{x}(t)}\Phi_{cb}(t) \qquad \Phi_{ab}(t_n) = \delta_{ab}, \tag{4.15}$$

along with the uncoupled dynamics (Eq. (4.14)) to produce

$$\Phi_{ab}(t_n + (k-1)\,\tau) = \frac{\partial\bar{x}_a(t_n + (k-1)\,\rho)}{\partial x_b(t_n)} \tag{4.16}$$

for $k = 1, \ldots, D_E$. This enables us to construct the Jacobian of the proxy vector with respect to $\mathbf{x}(t)$. However, we take $L = 1$, and this expression simplifies to

$$\frac{\partial\mathbf{S}(\mathbf{x})}{\partial\mathbf{x}(t_n)} = \begin{pmatrix} \Phi_{11}(t_n) & \cdots & \Phi_{1D}(t_n) \\ \Phi_{11}(t_n+\rho) & \cdots & \Phi_{1D}(t_n+\rho) \\ \vdots & \vdots & \vdots \\ \Phi_{11}(t_n+(D_E-1)\rho) & \cdots & \Phi_{1D}(t_n+(D_E-1)\rho) \end{pmatrix}. \tag{4.17}$$

The desired control term is then obtained by constructing the pseudoinverse of the singular value decomposition, which we address in a moment.

Starting from some initial condition, the control scheme proceeds through the observation window mapping back and forth between the physical and proxy space until the end of the observation window is reached. At each time t, Eq. (4.10) and (4.15) are integrated simultaneously to construct the proxy vector $\mathbf{S}(\mathbf{x}(t))$ and the Jacobian map $\partial\mathbf{S}(\mathbf{x})/\partial\mathbf{x}$. The pseudoinverse, $\partial\mathbf{x}/\partial\mathbf{S}$, is then evaluated to map the control coupling from time-delay space back to physical space. The resulting perturbation $\delta\mathbf{x}$ to the dynamics is then treated as a constant over the short time window $t \to t + \Delta t$, and a small step $\Delta t \leq \rho$ is made in physical space along the coupled dynamics (Eq. (4.27)). This procedure is then iterated at the next time point and continues in this way until the end of the observation window is reached.

When $D_E > 1$, this scheme generates control perturbations on *all* state components and, therefore, may be used to estimate *both* states *and* parameters by promoting the parameters to state variables with trivial dynamics $d\boldsymbol{\theta}/dt = 0$. This is in contrast to the ($D_E = 1$) synchronization control, which requires some other means to estimate the parameters. Most important, of course, the time-delay control technique allows one to extract more information from *existing* measurements, a feature that will be extremely important for many real applications, where extra measurements are either prohibitively expensive, time-consuming, or not technologically feasible.

The benefits of using time-delays will be discussed in further detail in the context of the numerical examples presented in subsequent sections. For the moment, however, we divert our attention to a technical matter that is of crucial importance. Namely, the calculation of the control term $\partial\mathbf{x}/\partial\mathbf{S}$ as a regularized local inverse.

4.5 Regularizing the Local Inverse of $\partial S/\partial x$

It is time to address the details of the computation of the pseudoinverse $\partial x/\partial S = (\partial S/\partial x)^+$. We wish to solve the linear system of equations,

$$\frac{\partial S}{\partial x}\delta x = \delta S, \qquad (4.18)$$

to determine the control perturbation in physical space $\delta x(t)$. We use that in the nudging formulation,

$$\frac{dx(t)}{dt} = F(x(t), \theta) + \delta x(t). \qquad (4.19)$$

This task may be formulated as an optimization problem that seeks to minimize a least squares objective function,

$$\left[\frac{\partial S}{\partial x}\cdot\delta x - \delta S\right]^2 = \sum_{l=1}^{LD_E}\sum_{b=1}^{D}\left|\frac{\partial S_l}{\partial x_b}(t)\,\delta x_b(t) - \delta S_l(t)\right|^2. \qquad (4.20)$$

In general, $\partial S(x)/\partial x$ is a $D_E L \times D$ rectangular matrix, and its inverse is not well posed. The system may be underdetermined or overdetermined depending on the value of D_E.

A common solution for ill-posed problems such as this is to include a regularization term in the objective function (Press et al. (2007); Miller (1970); Tikhonov and Arsenin (1977)),

$$\sum_{l=1}^{LD_E}\sum_{i=1}^{D}\left|\frac{\partial S_l}{\partial x_i}(t)\,\delta x_i(t) - \delta S_l(t)\right|^2 + \sum_{i,j=1}^{D}\left|\Gamma_{ij}(t)\,\delta x_j(t)\right|^2, \qquad (4.21)$$

for a suitably chosen $\Gamma(t)$. This process is known as Tikhonov-Miller regularization (Press et al. (2007); Miller (1970); Tikhonov and Arsenin (1977)). It allows us to choose $\Gamma(t)$ to give preference for particular solutions with desirable properties. Here we choose $\Gamma(t) = \sigma I$, where I is a $D \times D$ dimensional identity matrix, which in the limit $\sigma \to 0$ recovers the expression for the Moore-Penrose pseudoinverse. In addition to being arguably the simplest choice for $\Gamma(t)$, this form selects for solutions to the system (Eq. (4.18)) that minimizes the least squares norm of δx.

4.6 Computing the Pseudoinverse with Singular Value Decomposition

There are many numerical approaches available for constructing the pseudoinverse of an $m \times n$ matrix M. The simplest choice involves the direct inversion of the matrix product,

$$M^* = (M^T M)^{-1})\,M^T. \qquad (4.22)$$

However, this technique is known to incur numerical stability problems, which become especially problematic when $\mathbf{M}$ is ill-conditioned. The reason is that if $\mathbf{M}$ has condition number κ, then the product $\mathbf{M}^T\mathbf{M}$ will have condition number κ^2 and may be considerably more ill-conditioned than the matrix $\mathbf{M}$.

An alternative approach that does not suffer from such instability involves a singular value decomposition (SVD) of the matrix $\mathbf{M}$. The SVD can be thought of as a generalization of eigenvalue decomposition to non-square matrices. That is, it decomposes $\mathbf{M}$ into a product of three matrices,

$$\mathbf{M} = \mathbf{U}\mathbf{W}\mathbf{V}^{\dagger} \tag{4.23}$$

where $\mathbf{U}$ and $\mathbf{V}$ are unitary matrices of size $n \times n$ and $m \times m$, respectively, and $\mathbf{W}$ is an $m \times n$ rectangular diagonal matrix of singular values σ_i. The notation $\mathbf{V}^{\dagger}$ is the conjugate transpose of $\mathbf{V}$. The SVD is unique up to permutations and sign exchanges of the singular values. Most algorithms choose the singular values to be positive and ordered such that $\sigma_1 > \sigma_2 \ldots$.

Once the SVD is known, the pseudoinverse can be constructed as

$$\mathbf{M}^+ = \mathbf{V}\mathbf{W}^+\mathbf{U}^*, \tag{4.24}$$

where $\mathbf{W}^+$ is defined by taking the reciprocal of each non-zero element along the diagonal, leaving the zeros in place. In practice, however, only elements larger than some small tolerance are taken to be non-zero, while the others are replaced by zeros. This choice of tolerance determines the rank of the inverse, which plays a crucial role in the numeric stability of the algorithm and governs its overall performance.

We are now prepared to tackle examples illustrating how valuable the time-delay method can be, even when $L = 1$.

4.7 Lorenz96 Model

To illustrate the introduction of time delay coordinates, we turn to the Lorenz96 model (Lorenz (2006)), often used in the geophysical literature to illustrate ideas in SDA. We select $D = 20$.

The differential equations for the model are given by

$$\frac{dx_a(t)}{dt} = x_{a-1}(t)(x_{a+1}(t) - x_{a-2}(t)) - x_a(t) + f_a, \tag{4.25}$$

with $a = 1, 2, \ldots, D$, $x_{-1}(t) = x_{D-1}(t)$, $x_0(t) = x_D(t)$, and $x_1(t) = x_{D+1}(t)$; $D = 20$. Selecting the forcing parameter to be $f_a = f = 8.17$ for each index a, we generate chaotic orbits using a fourth order Runge-Kutta solver (Press et al. (2007)) with $\Delta t = 0.01$. From these data we select the first state variable $x_1(t)$ to serve as our data $y(t)$.

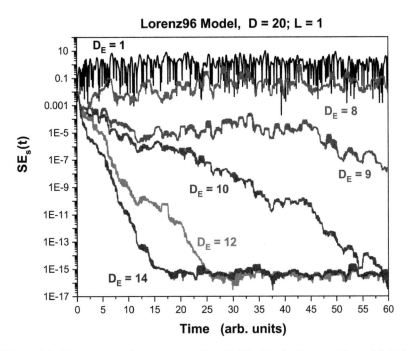

Figure 4.1 The synchronization error, Eq. (4.13), for the Lorenz96 model (4.25) with D = 20, and the same forcing in each component, $f_a = 8.17$. Information is passed from the data $Y_{1k}(t)$ to the model with $D_E = 1, 8, 9, 10, 12, 14$ using time delayed values of the observations of one state variable, here $x_1(t)$. As the dimension D_E of the time delay space is increased, more information is passed to the model, stabilizing the synchronization manifold.

From analysis of the Lorenz96 equations, we know that the value of L where instabilities on the synchronization manifold are 'cured' is approximately $L \approx 0.4D$ or, here, $L \approx 8$ (Kostuk (2012)) for $D = 20$. However, we have only the $L = 1 < 8$ measurement at present.

We treat the single parameter f as a 21st state variable $x_{21}(t)$ with the differential equation in physical space $dx_{21}(t)/dt = 0$. In the calculations on the Lorenz96 model we used $\tau = 0.1$, $t_n = 0 + n\Delta t$, and $g_{aa'} = 10$ (i.e. $g_{aa'}\Delta t = 1.0$).

4.8 Synchronization Errors in Time

In Fig. 4.1 we show $SE_s(t)$ for the Lorenz96 model with D = 20 for $D_E = 1, 8, 9, 10, 12, 14$. The model with data coupled into the dynamical equations has 21 dimensions. We show these synchronization errors over an observation window [0, 10] involving 1000 time steps of the Lorenz96 model with $\Delta t = 0.01$; $n_\rho = 10$. We anticipate that if D_E is large enough, this will approach zero rapidly. However, if D_E is too small, synchronization will not occur. Increasing D_E is equivalent

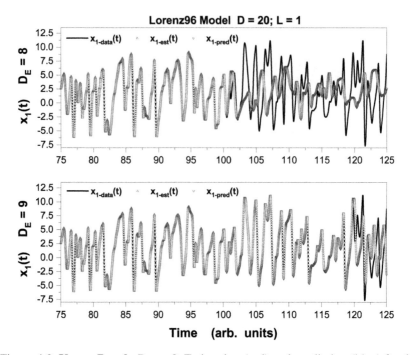

Figure 4.2 Upper Panel: $D_E = 8$. Estimation (red) and prediction (blue) for the Lorenz96 model with $D = 20$ for the physical space. The accuracy of the estimation and prediction of $x_1(t)$ is not good. Although the accuracy of the estimation for the observed physical state $x_1(t)$ looks quite good, one requires accurate estimation of the other 19 unobserved states and the parameter f to achieve good predictions. This is not accomplished at $D_E = 8$. The known 'data' values of $x_1(t)$ are shown as a black line. **Lower Panel:** Estimation (red) and prediction (blue) of the Lorenz96 model with $D = 20$ for the physical space and $D_E = 9$ for the time delay embedding space. The quality of the estimation (red) and prediction (blue) of the observed state $x_1(t)$ is quite good. The known values of $x_1(t)$ are shown as a black line. The prediction accuracy is limited by the chaotic behavior of this Lorenz96 model when $f = 8.17$. The deviation of the prediction near $t \approx 120$ is consistent with the largest positive conditional Lyapunov exponent of the model (Kostuk (2012)).

to increasing the number of measurements at each observation time τ_k and using this metric for identifying L_c. For the Lorenz96 model with $D = 20$ we found that, as D_E moves from below L_c for $D_E = 8$ to $D_E = 9$ and 10, synchronization occurs rapidly in time. This result illustrates quite clearly the distinction between the familiar nonlinear dynamics use of time-delay embedding to reconstruct the phase space of an observed system where D_E has to be larger than the attractor dimension D_A, which for the Lorenz96 model and our parameter choice is about $D_A \approx 12$. In using time delays to provide additional information to stabilize the synchronization manifold and allow accurate estimations, $D_E \approx 8$ is the dimension of the unstable synchronization manifold (Kostuk (2012)) for this version of the Lorenz96 model.

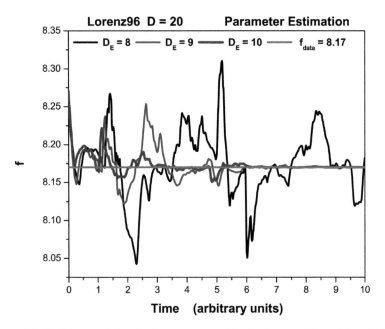

Figure 4.3 Estimates of the single parameter f in the Lorenz96 model with $D = 20$. The value used in generating the data is $f_{data} = 8.17$. Selecting $D_E = 8$ leads to inaccurate estimates of f. When $D_E = 9$ or 10, the parameter is very accurately estimated for $t \geq 6$. In these calculations, the parameter was taken as a state variable with the simple differential equation $\dot{f} = 0$.

More important for the validity of the model in **x** space is the ability to predict for $t > t_{final} = 100$. In the upper panel of Fig. 4.2 we show the known $x_1(t)$ (the data) in black, the estimated and predicted $x_1(t)$ in red and blue, respectively, for $D_E = 8$ and $D_E = 9$. For $D_E = 8$ the predictions beyond $= 100$ are not accurate, indicating that the estimates of all states at t_{final} and the estimate of the single parameter are not accurate. Note that the *estimate* for $x_1(t)$ appears rather good for $D_E = 8$ in $[t_0, t_{final}]$, indicating that the 'fit' to the data is not a useful metric for the quality of the estimation in the model. In the lower panel of Fig. 6.2 we display the known $x_1(t)$, the estimated $x_1(t)$, and the predicted $x_1(t)$ for $D_E = 9$. The quality of the prediction indicates that all state variables and f have been accurately estimated at $t_{final} = 100$.

In Fig. 4.3 we display the manner in which the estimation of the parameter f develops in time using time delay information. We show this for $D_E = 8$, 9, and 10. For $D_E = 8$ (black) the estimations of f are not accurate. For $D_E = 9$ and $D_E = 10$ the estimations of f become very accurate for $t \geq 6$. Treating the parameter as a state variable with differential equation $\dot{f} = 0$ successfully provides enough information using our procedure for large enough D_E.

Our next example uses the Lorenz96 model with $D = 10$, with different values for each of the forcings f_a in Eq. (4.25), as given in Table 4.1.

Table 4.1 *Forcing parameters f_a in the Lorenz96 model, $D = 10$. Actual values and estimations using time-delay data for $D_E = 1, 5, 6, 10$.*

Actual Value f_a	Estimated $D_E = 1$	Estimated $D_E = 5$	Estimated $D_E = 6$	Estimated $D_E = 10$
5.7	6.198	5.349	5.700	5.699
7.1	8.059	7.100	7.100	7.100
9.6	9.940	3.879	9.597	9.599
6.2	6.785	−2.439	6.204	6.200
7.5	7.723	4.569	7.495	7.499
8.4	9.151	13.463	8.403	8.400
5.3	5.555	−0.003	5.295	5.300
9.7	10.205	−0.261	9.702	9.699
8.5	9.199	−12.887	8.499	8.500
6.3	7.190	8.955	6.299	6.300

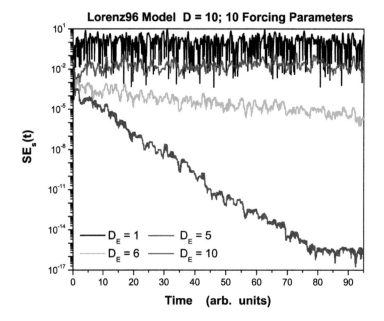

Figure 4.4 $SE_s(t)$ for the Lorenz96 model $D = 10$ and different forcing in each component with $D_E = 1, 5, 6, 10$. This shows that synchronization occurs with $D_E = 6$.

Figure 4.4 shows the temporal evolution of the synchronization error $SE_s(t)$ for different delay embedding dimensions D_E. While $D_E \leq 5$ is now sufficient for achieving synchronization, the simulation with $D_E = 6$ shows a slow convergence to zero, and the example using $D_E = 10$ exhibits a clear and fast transition to synchronization.

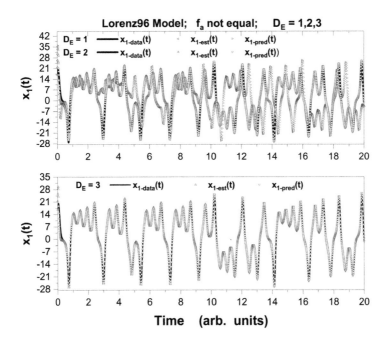

Figure 4.5 Data, Estimated, and Predicted values for $x_1(t)$ from the Lorenz96 model with $D = 10$ and different forcing f_a in each component of the model. For $D_E = 1$ and $D_E = 2$ the estimation and the prediction are not good nor is the model output synchronized to the data. For $D_E = 3$ we have excellent estimation and prediction. In Table 4.1 we display how the values of f_a improve as we increase D_E.

These results are confirmed in Fig. 4.5 where, in the top panel, synchronization and prediction fail for $D_E = 1$ and $D_E = 2$ but succeed for $D_E = 3$, as shown in the bottom panel.

The Lorenz96 model is an excellent testing ground for showing how one may utilize the information in the waveform of a time dependent measured state variable to provide the required additional information to stabilize an unstable synchronization manifold. Along this manifold information passes from data to a model of the nonlinear system producing that data. If one is able to make many measurements $x_1(t), x_2(t), ..x_L(t)$, where $L \leq$ some critical value, then time delays of each of the observed states allows the utilization of the information in the waveforms of the many available time series. If we have D_{Ei} time delays available in each of $i = 1, 2, ..L$ measurements, when $\sum_{i=1}^{L} D_{Ei}$ is large enough, the synchronization manifold will be stabilized, and accurate state and parameter estimations in a time window $[t_0, t_{final}]$ are possible. The example of the Lorenz96 model suggests that accurate predictions beyond t_{final} will then also be possible.

4.9 Rössler Hyperchaos

The four dimensional Rössler dynamical system provides another example for illustrating the application of time delay coordinates to problems of data assimilation.

The Rössler dynamics is described by the $D = 4$ ordinary differential equations:

$$\frac{dx_1(t)}{dt} = -x_2(t) - x_3(t)$$

$$\frac{dx_2(t)}{dt} = x_1(t) + p_1 x_2(t) + x_4(t)$$

$$\frac{dx_3(t)}{dt} = p_2 + x_1(t) x_3(t)$$

$$\frac{dx_4(t)}{dt} = p_3 x_3(t) + p_4 x_4(t). \tag{4.26}$$

We solved these starting from an arbitrary $\mathbf{x}(0)$. The parameters were selected to be $\{p_1, p_2, p_3, p_4\} = \{0.25, 3.0, -0.5, 0.05\}$.

A 'twin experiment' was conducted for this system. First, we generate a data set from these equations using the parameters as chosen. Starting from the initial condition $\mathbf{x}(0) = \{x_1(0), x_2(0), x_3(0), x_4(0)\} = \{-20, 0, 0, 15\}$, we generate four time series $\mathbf{x}_{data}(t)$. We used a time step $\Delta t = 0.025$ time units, and a fourth order Runge-Kutta integration protocol to generate the data.

These time series are stored as our data, and then $y(t) = x_{1-data}(t)$ is presented to our model as data. Thus, we have $L = 1$. The parameters are now treated as four additional state variables with dynamical equations for $\mathbf{p}(t) = \{p_1(t), p_2(t), p_3(t), p_4(t)\} = \{x_5(t), x_6(t), x_7(t), x_8(t)\}$; $d\mathbf{p}(t)/dt = 0$. It is through the coupling or control discussed above that the time dependence arises.

Next we form time delay data vectors $Y_k(t) = y(t + (k-1)\tau)$; $k = 1, 2, \ldots, D_E$ with $\rho = 4\Delta t$, and in the Rössler equations $dx_a(t)/dt = F_a(t)$; $a = 1, 2, .., 8$ we introduce the control term as

$$\frac{dx_a(t)}{dt} = F_a(\mathbf{x}(t)) + g \frac{\partial x_a(t)}{\partial S_k(\mathbf{x}(t))} (Y_k(t) - S_k(\mathbf{x}(t))), \tag{4.27}$$

where the time delay model output is formed from $x_1(t)$ and its time delays:

$$S_k(\mathbf{x}(t)) = x_1(t + (k-1)\rho); \quad k = 1, 2, \ldots, D_E. \tag{4.28}$$

We take the coupling strength to be the same here, $g = 10$, for all components of the Rössler system. $D_E = 8$. Just a reminder that $F_5(\mathbf{x})$, $F_6(\mathbf{x})$, $F_7(\mathbf{x})$, $F_8(\mathbf{x})$ are all zero.

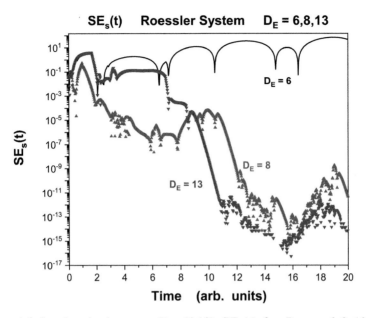

Figure 4.6 Synchronization error Eq. (4.13) $SE_s(t)$ for $D_E = 6, 8, 13$. The Rössler system was used with one measurement and using a proxy state vector of length D_E. One measurement, $x_1(t)$, was used; L = 1.

To find the inverse $\frac{\partial x_a(t)}{\partial S_k(\mathbf{X}(t))}$ we integrate the variational equation for the Rössler system as described in detail in earlier sections. The three unobserved initial model conditions for the model $\{x_2(0), x_3(0), x_4(0)\}$ were chosen at random from a uniform distribution that spans the region of state space visited by the system. In other words, we choose each $x_a(0)$ from a uniform distribution over the range $[x_a^{min}, x_a^{max}]$. For the sake of simplicity, we will report on one initial condition in particular, but the results presented are representative of a randomly chosen initial condition. The model output we generated with Eq. (4.27) comes from the randomly chosen initial state $x_{model}(0) = [-20, -18.6, 25.7, 122.4]$.

An observation window where $g \neq 0$ until $t_{final} = 20 = 800\Delta t$ is selected. For times after the observation window, we fix the model parameters at their estimated values $\mathbf{p}(t_{final})$, set $g = 0$ and integrate the model forward from $t_{final} = 20$ for another $8000\Delta t = 200$. In the observation window we selected $\{p_1(0), p_2(0), p_3(0), p_4(0)\} = \{0.125, 1.5, -0.25, 0.025\}$ as initial conditions, namely 50% of the known values.

During the integration we monitor the synchronization error

$$SE_s(t) = \frac{1}{2D_E} \sum_{k=1}^{D_E} (Y_k(t) - S_k(\mathbf{x}(t)))^2 \qquad (4.29)$$

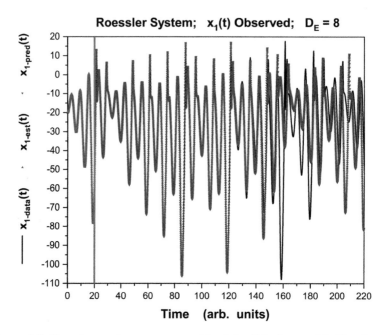

Figure 4.7 The observed component $x_1(t)$ of the Rössler model. The data was observed until $t_{final} = 20$, which is marked by the olive vertical line. The estimated $x_1(t)$ from the model with a control term using $D_E = 8$ is in red. After $t_{final} = 20$ a prediction $x_{1-pred}(t)$ (blue) was made by integrating the uncoupled $g = 0$ system forward for an additional 200 time steps. The data and the prediction deviate because of the chaotic behavior of the Rössler system.

as we proceed. At $t_{final} = 20$, g is set to zero, and using the state variables $\mathbf{x}(t_{final})$, we integrate the Rössler system forward for $t > t_{final}$. Of course, the 'states' $\{x_5(t), x_6(t), x_7(t), x_8(t)\}$ are constants for $t > t_{final}$: $\mathbf{p}(t) = \{p_1(t_{final}), p_2(t_{final}), p_3(t_{final}), p_4(t_{final})\}$.

In Fig. 4.6 we plot the synchronization errors $SE_s(t)$ for various values of D_E. We see that for $D_E = 6$ the synchronization error remains large, of order 10^{-1}, while as we use $D_E = 8$ and then 13, the synchronization error rapidly decreases in a significant way.

In Fig. 4.7 we display the observed state variable from the model output, $x_1(t)$, and we can see that the excellent predictions of $x_1(t)$ tell us we have estimated the states (and parameters) $\mathbf{x}(t_{final})$ quite well. The deviation of the predictions from the known data $y(t)$ is associated with the chaotic trajectories of the Rössler dynamics.

In a "twin experiment" we have information unavailable in an actual experiment, we can directly calculate what happens to the unmeasured components to check our claims. In Fig. 4.8 we display one of the unmeasured components $x_4(t)$ of the state

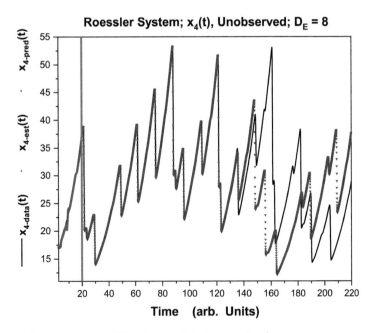

Figure 4.8 Comparison of the known data time series for an unobserved state variable $x_4(t)$ (black) of the Rössler system with the estimated $x_4(t)$ for $t \leq 20$. For $t > t_{final} = 20$ the predicted value of $x_4(t)$ is shown in blue. We see the prediction eventually fail because of the chaotic behavior of this Rössler system. We are able to compare this unobserved state variable with the known data, as this is a twin experiment.

of the Rössler system. It follows the known $x_{4-data}(t)$ generated at the outset of the twin experiment for the observation window [0, 20], then in the prediction window $t > 20$, $x_4(t)$ it continues to follow the data until the intrinsic chaotic instability separates the model output and data orbits as expected. Note that this separation happens at $t \approx 140$–150 with the prediction of the observed $x_1(t)$.

In a laboratory experiment we would be unable to assess the error in our parameter estimations, but a twin experiment allows us to explicitly illustrate that parameter estimation works using time-delay nudging (control) and with what accuracy. As shown in Fig. (4.9), while the parameter estimations may initially vary significantly, they soon settle on the correct values. In order to keep the integration stable while estimating parameters, maximum and minimum allowed search values for parameters were set at $p_{max} = 10$ and $p_{min} = -10$ for each. If the estimated parameter value exceeds p_{max} (p_{min}), that parameter is artificially fixed at the largest (smallest) allowed value until the estimate decreases. Numerical figures for the accuracy of parameter estimation are contained in Table 4.2. The values calculated are the fractional error of the parameters and states as estimated at time $t_{final} = 20$.

Table 4.2 *Errors in the estimated parameters for the Rössler dynamics:*
$\Delta p_a = \frac{p_a(t=20) - p_{a-data}}{p_{a-data}}$

D_E	Δp_1	Δp_2	Δp_3	Δp_4
6	29.7088	0.4368	1.1004	46.0390
7	32.4823	0.6324	1.0131	105.3668
8	1.8877e-11	4.1588e-9	4.1174e-8	4.7842e-10
9	3.9015	0.9653	1.0044	53.9609
10	3.8817	0.9681	7.1141	15.7934
11	1.9849e-11	1.6921e-10	8.1990e-09	1.8241e-10
12	26.8170	0.1683	1.8138	4.7644
13	1.3742e-12	4.9737e-12	3.8792e-10	9.8734e-12
14	1.1136e-13	1.0036e-13	2.3139e-11	7.7244e-13

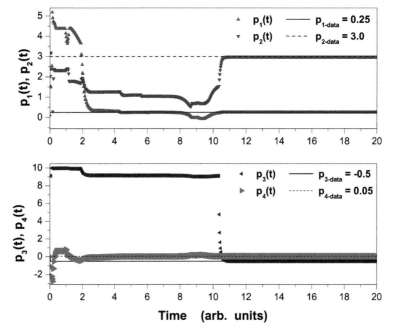

Figure 4.9 Estimation of the four parameters of the Rössler system over the observation interval. The bounded values of $p_3(t)$ are due to our limiting the variation of these quantities.

4.10 The Euler-Lagrange Equations for Measurement Terms in Proxy Vectors

In continuous time the action associated with Eq. (4.9) is

$$A(\mathbf{X}) = \int_{t_0}^{t_{final}} dt \, L(\mathbf{x}(t), \dot{\mathbf{x}}(t))$$

in which

$$L(\mathbf{x}(t), \dot{\mathbf{x}}(t)) = \sum_{l=1}^{L} \sum_{k=1}^{D_E} \frac{R_m(t)}{2} \left\{ Y_{lk}(t) - S_{lk}(t) \right\}^2$$

$$+ \sum_{a=1}^{D} \frac{R_f^{(x_a)}}{2} \left[\frac{dx_a(t)}{dt} - F_a(\mathbf{x}(t), \boldsymbol{\theta}) \right]^2. \tag{4.30}$$

The Euler-Lagrange equations for the extremum of this action is

$$\frac{d^2 x_a(t)}{dt^2} = \Omega_{ab}(\mathbf{x}(t)) \frac{dx_b(t)}{dt} +$$

$$+ \frac{1}{2} \frac{\partial [\mathbf{F}(\mathbf{x}(t)\boldsymbol{\theta})]^2}{\partial x_a(t)} + \frac{1}{2R_f^{(x_a)}} \frac{\partial \nabla \cdot \mathbf{F}(\mathbf{x}(t))}{\partial x_a(t)}$$

$$+ \frac{R_m(t)}{R_f^{(x_a)}} \frac{\partial S_{lb}(\mathbf{x}(t))}{\partial x_a(t)} \left\{ S_{lb}(\mathbf{x}(t)) - Y_{lb}(t) \right\}, \tag{4.31}$$

with, as earlier,

$$\Omega_{ab}(\mathbf{x}(t)) = \frac{\partial F_a(\mathbf{x}(t), \boldsymbol{\theta})}{\partial x_b(t)} - \frac{\partial F_b(\mathbf{x}(t), \boldsymbol{\theta})}{\partial x_a(t)}. \tag{4.32}$$

We see that the measurement error term has changed to reflect the fact we use D_E time delays in the comparison of the output model state variables $\mathbf{x}(n)$, with their own time delays, and the data, with its time delayed versions. The model error term keeps its form and moves the stochastic model forward from t_n to t_{n+1}. We labeled this EL equation as new-nudging.

4.11 Simplified Use of Waveforms

One of the features of the proxy state method is the coupling of nudging terms in the state variables to an effective nudging in the parameter equations. Absent nudging, the parameters satisfy

$$\frac{d\boldsymbol{\theta}(t)}{dt} = 0. \tag{4.33}$$

When the matrices coming from nudging are introduced, this is promoted to

$$\frac{d\boldsymbol{\theta}(t)}{dt} = g \frac{\partial x_a(t)}{\partial S_k(\mathbf{x}(t))} (Y_k(t) - S_k(\mathbf{x}(t))), \tag{4.34}$$

adding an effective nudging to the parameters as well.

In a simplification of the method derived from an action, as developed here, Pazó et al. (2016) replaced the state space structure in Eq. (4.34) with an "old"-nudging matrix $\mathbf{g}$ that achieves the connection of a proxy vector to the parameters $\boldsymbol{\theta}$ directly:

$$\frac{d\boldsymbol{\theta}(t)}{dt} = g_{\boldsymbol{\theta},k}(Y_k(t) - S_k(\mathbf{x}(t))). \tag{4.35}$$

Somehow, the need to remove the coupling at the end of the calculation was ignored; nonetheless, accurate estimates of parameters and state variables were accomplished.

5

Annealing in the Model Precision R_f

In Chapter 1 we noted that it is (really) hard to find the global minimum of a nonlinear function of $\mathbf{X}$ such as our standard action

$$A(\mathbf{X}) = \sum_{n=0}^{N} \frac{R_m(n)}{2} \sum_{l=1}^{L} \{x_l(n) - y_l(n)\}^2$$
$$+ \sum_{n=0}^{N-1} \sum_{a=1}^{D} \frac{R_f^{(x_a)}}{2} \{x_a(n+1) - f_a(\mathbf{x}(n), \boldsymbol{\theta})\}^2. \tag{5.1}$$

At the same time, the global minimum action associated with the path $\mathbf{X}^0$ is the optimum choice for using Laplace's method to estimate expected values of the conditional probability distribution $P(\mathbf{X}|\mathbf{Y})$, so it is of more than passing interest to seek methods to find $\mathbf{X}^0$.

This is a dilemma that has drawn the attention, for example, of meteorologists (Pires et al. (1996)) wishing to make accurate numerical weather predictions using variational principles to determine $\mathbf{X}^0$. It is also likely that others doing data assimilation here or there face this issue, but perhaps proceed unaware of the challenge.

There is a vivid illustration of how the action develops local minima in Miller et al. (1994) with an insightful discussion of the issues in the text near Fig. (6.1) of the comprehensive book by Evensen (2009).

This chapter takes a different look at this question.

The culprit making the goal of locating the global minimum in the standard model action a hard task is the model error term, which contains the nonlinear function $\mathbf{f}(\mathbf{x}(n), \boldsymbol{\theta})$:

$$\sum_{n=0}^{N-1} \sum_{a=1}^{D} \frac{R_f^{(x_a)}}{2N} \{x_a(n+1) - f_a(\mathbf{x}(n), \boldsymbol{\theta})\}^2. \tag{5.2}$$

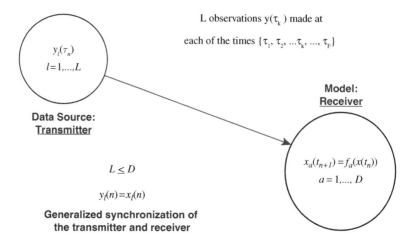

L observations $y(\tau_k)$ made at

each of the times $\{\tau_1, \tau_2, ...\tau_k, ..., \tau_F\}$

$y_l(\tau_n)$

$l = 1,...,L$

Model:
Receiver

Data Source:
Transmitter

$L \leq D$

$x_a(t_{n+1}) = f_a(x(t_n))$

$a = 1,..., D$

$y_l(n) = x_l(n)$

Generalized synchronization of
the transmitter and receiver

Figure 5.1 A graphical display of the nonlinear dynamics associated with data assimilation. At times τ_k L-dimensional noisy data vectors $\mathbf{y}(\tau_k)$ are measured. These measurements contain information about the dynamical system being observed. The data are presented to the model system $\mathbf{x}(t_{n+1}) = \mathbf{f}(\mathbf{x}(t_n), \boldsymbol{\theta})$, selected by the user to capture the dynamical time development of the D-dimensional state $\mathbf{x}(t)$ of the system. $D \geq L$; often $D \gg L$. The system is observed over a temporal observation window $[t_0, t_{final}]$. The measurement error term of the action $A(\mathbf{X}) = -\log[P(\mathbf{X})]$ works to move the model state to the data: $x_l(\tau_k) \approx y_l(\tau_k)$; $l = 1, 2, \ldots, L$, subject to satisfying the model. This matching is generalized synchronization of the data and the model. In nonlinear systems there are often intrinsic instabilities on the **synchronization manifold**, where $x_l(\tau_k) = y_l(\tau_k)$. These interfere with the matching effort and must be controlled by improving the model and/or increasing the number of measurements at L at each observation time.

The form of the model error term was chosen to capture the many ways noise or errors in the model might spoil the deterministic model equations $x_a(n + 1) = f_a(\mathbf{x}(n), \boldsymbol{\theta})$, which become accurate as $R_f^{(x_a)} \rightarrow \infty$. That turned our focus on the limit where the precision of the model is important.

In this chapter we have a look at how the opposite limit $R_f^{(x_a)} \rightarrow 0$ can play a role in locating the global minimum of $A(\mathbf{X})$ and how tracking $R_f^{(x_a)}$ along the way from 0 to ∞ may assist in understanding how the variational assimilation process operates.

The utilization of these ideas in Monte Carlo sampling of the conditional probability distribution $P(\mathbf{X}|\mathbf{Y})$ will be taken up in Chapter 7.

R_f is called a *hyperparameter* of the data assimilation formulation. Its structure and magnitude are not set by physical considerations.

R_m, setting the scale for the measurement error magnitude, is quite different, as it is connected with properties of the measurement devices used in collecting data. In the metadata we touched on at the outset of these considerations, the properties of R_m were identified as important and quite likely known. In the caption to Fig. 5.1

we refer to generalized synchronization (Sushchik et al. (1995); Abarbanel et al. (1996); Parlitz et al. (1996); Tang et al. (1998)), which is synchronization between two *unlike* signals. Here that refers to the equality of the noisy observations $y_l(\tau_k)$ and the model output $x_l(\tau_k)$. This equality depends on the conditional Lyapunov exponents associated with directions transverse to the synchronization manifold being less than zero (Pecora and Carroll (1990)).

5.1 Varying the Hyperparameter R_f

5.1.1 $R_f \to 0$

In the standard action Eq. (5.1) we may vary R_f as we wish, and, if we take $R_f \to 0$, we see that the action is now quadratic in the components of $\mathbf{X}$. It is also enormously degenerate as only the L (out of D) state variables are present. Its minima are easy to find.

The $R_f = 0$ action is now

$$A_{R_f=0}(\mathbf{X}) = \sum_{n=0}^{N} \sum_{l=1}^{L} \frac{R_m(n)}{2N} [x_l(n) - y_l(n)]^2, \tag{5.3}$$

and, as usual, $R_m(n)$ vanishes except at times $t_n = \tau_k$, where measurements are performed.

We want to explore the determination of the minima of actions starting at $R_f = 0$ and slowly increasing R_f until it is quite substantial. We'll begin at $R_f = 0$ and identify an ensemble of N_I global minimum paths, then we schedule increases in R_f and use the N_I extremum paths from one setting of R_f as starting locations for the next setting.

We select a collection, one might call it an ensemble, of initial paths $\mathbf{X}_{init}^q$; $q = 1, 2, \ldots, N_I$ satisfying

$$\left. \frac{\partial A_{R_f=0}(\mathbf{X})}{\partial \mathbf{X}} \right|_{\mathbf{X}_{init}^q} = 0, \tag{5.4}$$

each of which is a (degenerate) global minimum (Ye et al. (2015a)) of the $R_f = 0$ action. We will then schedule increases in R_f, moving away from $R_f = 0$, and track these N_I minimum action paths as R_f becomes larger ($R_f \geq 0$).

5.1.2 Initial Paths

To create our choice of initial paths, we integrate the model forward through N discrete times $t_0 + n\Delta t$; $n = 0, 1, 2, \ldots, N$ to $t_{final} = t_N$ starting with some $\mathbf{x}(0)$. $\mathbf{x}(0)$ is initialized such that $x_l(0) = y_l(0)$, if a measurement is available at $t_0 = 0$. If there is no measurement at t_0, $x_a(0)$ is drawn from a uniform distribution covering the dynamical range of the model for $a = 1, \ldots, D$. The parameters $\boldsymbol{\theta}$ are also drawn from an appropriate uniform distribution. This is done N_I times using N_I selections of $x_a(0)$ and $\boldsymbol{\theta}$. A broad range of initial choices for $\boldsymbol{\theta}$ should be made consistent with any conditions known about them.

The model is then integrated forward in time using steps of size Δt; $t_n = t_0 + n\Delta t$, with the observed values $y_l(\tau_k)$ replacing $x_l(t_n = \tau_k)$ at each time we have a measurement, $t_k = \tau_k = n_k \Delta t$ for $k = 1, \ldots, F$:

$$x_a(n) = y_a(n), \quad a = 1, \ldots, L,$$
$$x_a(n + 1) = f_a(\mathbf{x}(n), \boldsymbol{\theta}), \quad a = L + 1, \ldots, D. \tag{5.5}$$

This completes our construction of N_I initial paths $\mathbf{X}_{init}^q$. The freedom in choosing $\mathbf{x}(0)$ and $\boldsymbol{\theta}$ gives us flexibility to generate multiple such initial paths (an ensemble of initial paths) $\mathbf{X}_{init}^q$, $q = 1, \ldots, N_I$ at $R_f = 0$. These are retained for future use.

Precision Annealing is the idea that if we choose any of the initial paths $\mathbf{X}_{init}^q$ to use as an ensemble member for a variational optimization (Quinn (2010); Ye (2016); Ye et al. (2015a,b)) at the global minimum for $R_f = 0$, as we slowly increase R_f, we will stay in a region of path space where our paths will arrive at the smallest minimum of the action for $R_f \neq 0$, for a variational calculation. In the following we discuss locating minima of an action $A_{R_f=R_{f0}\alpha^\beta}(\mathbf{X})$.

5.1.3 $R_f = R_{f0} > 0$

We adopt the following annealing schedule,

$$R_f = R_{f0}\alpha^\beta, \tag{5.6}$$

with $\alpha > 1$ and $\beta = 0, 1, 2, \ldots, \beta_{max}$. At each value of β we perform N_I calculations of some numerical nonlinear minimization operation.

At $\beta = 0$, we choose $R_f = R_{f0} = $ **a very small number**. Then we begin the nonlinear optimization at the N_I distinct initial paths $\mathbf{X}_{init}^q$ as we steadily increase β. A choice of α near unity leads to the slow increase in R_f as β increases, introducing the nonlinearity of the model in an adiabatic manner.

At this stage one must choose a numerical nonlinear optimization strategy to find the minima of the action when $R_f > 0$. We have found the algorithms contained in the public domain IPOPT programs (Wachter and Biegler (2006)) to be accurate and remarkably efficient. We use it from now on without further comment.

At each R_f (equivalently, at each β) value, N_I numerical optimizations are performed starting from the solution generated by the procedure at the previous R_f (or β). α is fixed in this procedure.

When the numerical optimization calculations at $\beta = 0$; $R_f = R_{f0}\alpha^0$ are completed, we will have $q = 1, 2, \ldots N_I$ slightly different resulting paths $\mathbf{X}_0^q$, each of which gives a minimum of the action

$$A_{R_f = R_{f0}\alpha^0}(\mathbf{X}) = A_0(\mathbf{X}) = \sum_{n=0}^{N} \sum_{l=1}^{L} \frac{R_m(n)}{2} [x_l(n) - y_l(n)]^2$$

$$+ \sum_{n=0}^{N-1} \frac{R_{f0}^{(x_a)} \alpha^0}{2} [x_a(n+1) - f_a(\mathbf{x}(n), \boldsymbol{\theta})]^2, \tag{5.7}$$

and satisfies

$$\left. \frac{\partial A_0(\mathbf{X})}{\partial \mathbf{X}} \right|_{\mathbf{X}_0^q} = 0. \tag{5.8}$$

5.1.4 $R_f = R_{f0}\alpha^\beta$; $\beta \geq 1$

Next we move $R_f \to R_{f0}\alpha^1$ and repeat the N_I optimization calculations starting now with the initial paths $\mathbf{X}_0^q$ and resulting in N_I new paths $\mathbf{X}_1^q$ at $R_f = R_{f0}\alpha^1$. Each of the paths $\mathbf{X}_1^q$ minimizes the action

$$A_{R_f = R_{f0}\alpha^1}(\mathbf{X}) = A_1(\mathbf{X}) = \sum_{n=0}^{N} \sum_{l=1}^{L} \frac{R_m(n)}{2} [x_l(n) - y_l(n)]^2$$

$$+ \sum_{n=0}^{N-1} \frac{R_{f0}^{(x_a)} \alpha^1}{2} [x_a(n+1) - f_a(\mathbf{x}(n), \boldsymbol{\theta})]^2, \tag{5.9}$$

satisfying

$$\left. \frac{\partial A_1(\mathbf{X})}{\partial \mathbf{X}} \right|_{\mathbf{X}_1^q} = 0. \tag{5.10}$$

Continuing on, we move R_f to $R_{f0}\alpha^2$ and starting with the ensemble of N_I paths $\mathbf{X}_1^q$, we minimize

$$A_{R_f = R_{f0}\alpha^2}(\mathbf{X}) = A_2(\mathbf{X}) = \sum_{n=0}^{N} \sum_{l=1}^{L} \frac{R_m(n)}{2} [x_l(n) - y_l(n)]^2$$

$$+ \sum_{n=0}^{N-1} \frac{R_{f0}^{(x_a)} \alpha^2}{2} [x_a(n+1) - f_a(\mathbf{x}(n), \boldsymbol{\theta})]^2, \tag{5.11}$$

satisfying

$$\left. \frac{\partial A_2(\mathbf{X})}{\partial \mathbf{X}} \right|_{\mathbf{X}_2^q} = 0, \tag{5.12}$$

resulting in N_I new paths $\mathbf{X}_2^q$.

We repeat this calculation, increasing β at each step by unity $\beta \to \beta + 1$, eventually reaching $\beta = \beta_{max} - 1$. Minimizing the action

$$A_{R_f=R_{f0\alpha}(\beta_{max}-1)}(\mathbf{X}) = A_{(\beta_{max}-1)}(\mathbf{X}) = \sum_{n=0}^{N}\sum_{l=1}^{L}\frac{R_m(n)}{2}[x_l(n) - y_l(n)]^2$$

$$+ \sum_{n=0}^{N-1}\frac{R_{f0}^{(x_a)}\alpha(\beta_{max}-1)}{2}[x_a(n+1) - f_a(\mathbf{x}(n), \boldsymbol{\theta})]^2, \tag{5.13}$$

satisfying

$$\left.\frac{\partial A_{(\beta_{max}-1)}(\mathbf{X})}{\partial \mathbf{X}}\right|_{\mathbf{X}^q_{(\beta_{max}-1)}} = 0, \tag{5.14}$$

results in N_I new paths $\mathbf{X}^q_{(\beta_{max}-1)}$.

As a final step we use the paths $\mathbf{X}^q_{(\beta_{max}-1)}$ to initialize the minimization of

$$A_{R_f=R_{f0\alpha}\beta_{max}}(\mathbf{X}) = A_{\beta_{max}}(\mathbf{X}) = \sum_{n=0}^{N}\sum_{l=1}^{L}\frac{R_m(n)}{2}[x_l(n) - y_l(n)]^2$$

$$+ \sum_{n=0}^{N-1}\frac{R_{f0}^{(x_a)}\alpha\beta_{max}}{2}[x_a(n+1) - f_a(\mathbf{x}(n), \boldsymbol{\theta})]^2, \tag{5.15}$$

satisfying

$$\left.\frac{\partial A_{\beta_{max}}(\mathbf{X})}{\partial \mathbf{X}}\right|_{\mathbf{X}^q_{\beta_{max}}} = 0. \tag{5.16}$$

When reaching $\beta = \beta_{max}$, we end this sequence of operations. Along the way we have created sets of N_I paths $\mathbf{X}^q_\beta$ at each β and evaluated the action $A(\mathbf{X}^q_\beta)$. Plotting the $A(\mathbf{X}^q_\beta)$ versus β is of interest, and we do so below.

5.1.5 How Does One Select β_{max}?

There are two questions here:

- The first question is this: at each $\beta = 0, 1, 2 \ldots, \beta_{max}$, how does one select the number of measurements L made at each time τ_k when observations are accomplished?
- The second question is this: when does the information in the data become properly represented in the selected model?

Question (1): As we increase L the information delivered about the underlying nonlinear processes increases. When all of the unstable directions associated with the synchronization manifold have been probed and through

$\sum_{n=0}^{N} \frac{R_m(n)}{2} \sum_{l=1}^{L} \{(x_l(n) - y_l(n)\}^2$, perturbations off that manifold in state space will have been confined in a harmonic oscillator well and can be controlled. This is explored in a quantitative manner by the evaluation of the *conditional Lyapunov exponents* (Pecora and Carroll (1990); Kostuk (2012)). As we increase L, there is a critical value where they all become negative; this is the definitive sign of the arrival of stability. For L larger than the critical value, the possibility of a β_{max} arises.

This synchronization is an example of **generalized synchronization** (Sushchik et al. (1995); Abarbanel et al. (1996); Parlitz et al. (1996); Tang et al. (1998)) in which signals from nonidentical sources may synchronize the two sources. Synchronization of two signals, here $\mathbf{y}(\tau_k)$ and $\mathbf{x}(\tau_k)$, is possible when the conditional Lyapunov exponents are negative, and that needs to happen here. If any of the conditional Lyapunov exponents are positive, then the optimization algorithms, as they search through path space making perturbations here and there, will move off the synchronization manifold and grow exponentially rapidly away from that manifold. The paths may return subsequently to the neighborhood of the synchronization manifold, but, as it is still unstable, move away again. All this activity near $x_l(\tau_k) \approx y_l(\tau_k)$ leads to uncertainty or chaotic activity in path space which manifests itself in the action level plots with multiple local minima.

Question (2): Once L is large enough, the consistency of the selected model with the data is explored when R_f increases. If the model captures the dynamics being observed, the model error term will decrease rapidly, and the model error term may decrease fast enough so the action becomes effectively independent of R_f and levels out to track the measurement error term, which is not dependent on R_f.

An ancillary matter can also be important. If the dynamics $\mathbf{x}(n+1) = \mathbf{f}(\mathbf{x}(n), \boldsymbol{\theta})$ yields chaotic trajectories, then when the time between one measurement time and the next is larger than the inverse of the largest global Lyapunov exponent, the chaos that arises because the dynamics is uncontrolled over that long time interferes with accurate searches of paths that are minima of the action. This is some of what one is seeing in Fig. 6.1 of Evensen (2009). The cure proposed for this (Evensen (2009)) is to make the measurements more frequently. This suggestion begs the question of understanding the nonlinear dynamics of the SDA problem.

We now turn to two twin experiments which contain instructive models that illustrate these remarks.

5.2 Lorenz96 Model with $D = 5$

We first return to the dynamical equations introduced by Lorenz (2006):

$$\frac{dx_a(t)}{dt} = x_{a-1}(t)(x_{a+1}(t) - x_{a-2}(t)) - x_a(t) + f_a \qquad (5.17)$$

and $a = 1, 2, \ldots, D$; $x_{-1}(t) = x_{D-1}(t)$; $x_0(t) = x_D(t)$; $x_1(t) = x_{D+1}(t)$. The f_a are fixed parameters. The equations for the states $x_a(t)$; $a = 1, 2, \ldots, D$ are meant to describe 'stations' on a periodic spatial lattice. The dynamical equations are easily scaled in their overall dimension D, in the number of forcing parameters, and in twin experiments in data assimilation in the number of measurements L made at each τ_k. So these dynamics have become a common testbed for ideas in data assimilation.

We perform a twin experiment wherein we generate D time series using an adaptive fourth order Runge-Kutta algorithm (Press et al. (2007)) with a time step $\Delta t = 0.025$. To these we add Gaussian noise with mean zero and variance $\sigma^2 = 0.25$ to each time series $x_a(t)$. These noisy versions of our model time series constitute our 'data.' We selected $f_a = f = 8.17$.

Independence of $A_0(\mathbf{X}^q)$ from R_f indicates that the model output has matched the deterministic dynamics $\mathbf{x}(n + 1) = \mathbf{f}(x(n), \boldsymbol{\theta})$ quite well. The remaining term in the action is then

$$A_{\beta=0}(X) = \sum_{n=0}^{M} \sum_{l=1}^{L} \frac{R_m(l, n)}{2} [x_l(n) - y_l(n)]^2. \tag{5.18}$$

As the values $[y_l(n) - x_l(n)]$ are distributed as $\mathcal{N}(0, \sigma^2)$ by our choice, the measurement error term $\sum_{n=0}^{M} \sum_{l=1}^{L} [(x_l(n) - y_l(n))/\sigma]^2/2$ has a χ^2-distribution with $L(M + 1)$ degrees of freedom (Feller (2008)). The mean and RMS variation of this distribution over different choices of noise waveforms are $(M + 1)L/2$ and $\sqrt{(M + 1)L/2}$, respectively. This expected measurement error level is shown in our action value versus R_f plots by a heavy horizontal line. When the action levels as a function of R_f reach this expected χ^2 lower limit, we have a path $\mathbf{X}^0$ on which the model behavior is consistent with the data within the noise level of the data.

The measurement window is from $t_0 = 0$ to $t_N = t_f = 4.0$, so $N = 160$. L 'measurements' are made at each time step; these are the $\mathbf{y}(t_n)$. The measurement error matrix $\mathbf{R}_m$ is taken to have diagonal elements at each measurement time t_n and is zero at other times. Its magnitude is $R_m = 1/\sigma^2 = 4$. The model error matrix is also taken as diagonal, with elements along the diagonal $R_f = R_{f0}2^\beta$, and we take $\alpha = 2$, $\beta = 0, 1, 2, \ldots, 22$. $R_{f0} = 0.01$.

We begin with one measurement $y_1(n)$ among the five possible states, and in Fig. 5.2 we display the $\log_{10}$ of the action $A_0(\mathbf{X})$ evaluated at each of the $N_0 = 100$ saddle paths for $R_f = R_{f0}2^\beta$ with $R_{f0} = 0.01$. We begin with $\beta = 0$ and increase it to $\beta = 22$.

In using IPOPT, we provided the gradient of $A_0(\mathbf{X})$ in an analytical form to the algorithm. We initialized the search at $R_f = R_{f0}$ with N_I initial paths

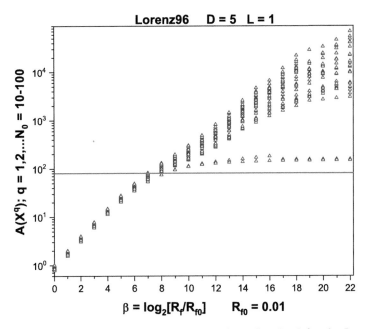

Figure 5.2 Action levels as a function of $\beta = \log_2[R_f/R_{f0}]$ for the Lorenz96 model, Eq. (5.17), with $D = 5$, $L = 1$, $R_{f0} = 0.01$. As R_f increases, the model error decreases. The horizontal line shows the expected value of the measurement error terms in the action. This term is distributed as χ^2 with expected value $(M + 1)L/2 = 80.5$ and RMS error $\sqrt{(M + 1)L/2} \approx 9$. The resulting saddle paths all have action levels above the χ^2 expected value for the measurement error action alone. This is an indication that $L = 1$ is not sufficient for identifying a good path for the Lorenz96 model.

$\mathbf{X} = \{\mathbf{x}(0), \mathbf{x}(1), \ldots, \mathbf{x}(m)\}$ as described above. At $R_f = R_{f0}$ we selected the unobserved states at each time step from a uniform distribution in the interval $[-10, 10]$. This is approximately the dynamic range of state variables in the Lorenz96 model. At fixed R_m each search procedure as we slowly increase R_f yields N_I saddle paths $\mathbf{X}^q$ and associated action levels $A_0(\mathbf{X}^q)$. As R_f increases many initial paths may lead to the same action level.

In Fig. 5.2 we display the action values associated with minima in $A(\mathbf{X})$ produced by our ensemble of N_I initial paths as a function of R_f for $D = 5$ and $L = 1$. Clearly there are many local minima, and none of the minima at large R_f satisfies the criterion that the model error becomes small enough that the action level is dominated by the measurement error term. This is an indication that $L = 1$ is not sufficient to remove the multiple minima arising from instabilities on the synchronization manifold.

As one can see in Fig. 5.2, the degenerate action levels at $R_f = 0$ are split at $\beta = 0$ and then rise until, around $\beta = 12$; $R_f \approx 100$, two levels split off from

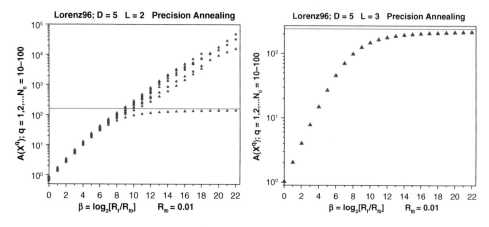

Figure 5.3 **Left Panel**: Action Levels as a function of R_f for the Lorenz96 model, D = 5, L = 2, R_{f0} = 0.01. We used $y_1(t)$ and $y_3(t)$ as data in the action. **Right Panel**: Action Levels as a function of R_f for the Lorenz96 model, D = 5, L = 3, R_{f0} = 0.01. We used $y_1(t)$, $y_3(t)$, and $y_5(t)$ as data in the action.

the rest and become somewhat independent of R_f. There are still two quite close levels. Also shown is the expected value of the χ^2-distributed measurement error at 80.5. The distance of the action levels of the paths giving $A_0(X)$ near values about 150 tell us that these paths are unlikely to give consistency of the model with the data. Using either of these two paths to give us the full model states at the end of the estimation window $t_{final} = 4$ to predict beyond t_{final} gives quite inaccurate predictions.

The many local minima of the action tend to grow as R_f, indicating the model error term for those paths in the action that has an explicit coefficient R_f are not satisfying the Lorenz96 model equations very well.

Having learned that $L = 1$ is not sufficient to capture the information in the data, we turn to $L = 2$. We present $L = 2$ *noisy* measurements, $y_1(n)$ and $y_3(n)$, to the model and again evaluate the action levels along the paths that are extrema of the action as we vary β. Each path has $(N + 1)D = 805$ components, so the annealing problem is a search for saddle paths of the action Eq. (5.1) in an 805-dimensional space. We take the distribution of the three unobserved states at t_0, the beginning of the observation window, to be uniform over the dynamical range of $\mathbf{x}(t_0)$.

At $\beta = 0$ the degenerate action levels from $R_f = 0$ are split slightly. We follow these to larger values of R_f (Fig. 5.3). At low R_f the resolution in path space is very coarse, and our search is successful for finding low lying action levels. Figure 5.3 shows quite clearly that there are many paths with similar action level until we reach $\beta \approx 12$, and after that only one remains independent of R_f as the other action levels rise. The expected value of the χ^2-distribution of measurement errors in the action is $N_{data} R_m \sigma^2/2$. This is 161 here, and it is shown in the figure

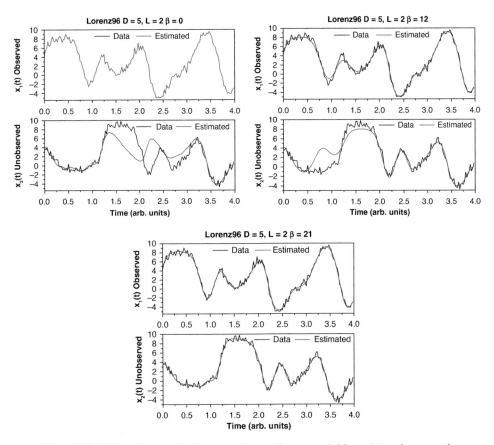

Figure 5.4 Estimation results for one observed state variable $x_1(t)$ and one unobserved state variable $x_2(t)$ during the annealing procedure. **Left Panel**: $\beta = 0$. **Right Panel**: As β increases to 12, the estimated observed state becomes smoother. **Bottom Panel**: The observed and unobserved state variables arrive at their true states when β is large enough. In this display $\beta = 21$.

as a heavy horizontal line (olive). The action level for $\mathbf{X}^0$ is very near this χ^2 consistency condition, suggesting that the path $\mathbf{X}^0$ expresses consistency of the model and the data.

When we increase the number of measured time series to $L = 3$, the results in Fig. 5.3 show that one path alone emerges from the degeneracy at $R_f = 0$ and after $\beta \approx 12$ is again nearly independent of R_f and close to the expected limit from the χ^2 distribution.

To get some insight into how the annealing procedure proceeds in the sequence of estimates for the observed and unobserved states of the model to which $L = 2$ 'data' time series are presented, we show in Fig. 5.4 the estimated and the 'data' time courses for both an observed state variable $x_1(t)$ and an unobserved state variable $x_2(t)$ for selected values of $\beta = 0$, 12 and $\beta = 21$.

In Fig. 5.4 representative time series that are part of the path for different values of β in the lowest action level are plotted to illustrate the annealing process in detail. For very small β, say 0, $R_f = R_{f0} = 0.01$, the top two panels of Fig. 5.4 show the known and estimated components $x_1(t)$, observed, and $x_2(t)$, unobserved, from one of the saddle paths. Since $R_m \gg R_f$, and the measurement error dominates the overall size of the action, paths are forced to follow the noisy measurements almost exactly so as to minimize the measurement error (i.e. $x_l(t) \approx y_l(t)$). The effect of the model error term is quite small with $R_f = 0.01$; the unmeasured states are usually undetermined. Its form depends on the initial random guess path. In this example, the initial path happens to be chosen near the true path, and the unobserved state x_2 is close to the known data at the beginning and end of the window.

As β is increased to 12, $R_f \approx 40$, we have moved from a regime where R_f is quite small to a regime where R_f has become sizable. The role of the model error is no longer insignificant. The trajectory of the observed state $x_1(t)$ is smoother, passing through the middle of the noise fluctuations, but not tracking the noise as was done at $\beta = 0$. The greater R_f, the more information is input from the model. This information from the model helps the observed state filter out the noise to some extent. When β increases up to 21, $R_f \gg R_m$, it enforces the model more and more exactly, $x_a(n + 1) \approx f_a(\mathbf{x}(n))$. Both observed states and unobserved states converge to the true path for the lowest action level. The size of action $A_0(\mathbf{X})$ matches the observation error residual $N_{data} R_m \sigma^2/2$.

It is important to note that if we begin our search for the saddle paths $\mathbf{X}^q$ at large values of R_f, we are almost sure to miss the actual path $\mathbf{X}^0$ that gives the lowest action level, since the Hessian matrix of $A_0(\mathbf{X})$ is ill-conditioned when R_f is large and the lowest action level occupies a tiny corner of the large (here 805 dimensional) path space. See Fig. 4.6 in Quinn (2010).

5.3 Hodgkin-Huxley NaKL Neuron

Our second example is chosen from neurobiology (Johnston and Wu (1995); Sterratt et al. (2011)). We selected a fairly standard Hodgkin-Huxley (HH) neuron model consisting of four state variables: the voltage $V(t)$ across the cell membrane as well as three voltage dependent gating variables for Na^+ and K^+ channels $m(t), h(t)$, and $n(t)$. The equation governing changes in voltage across the cell membrane is current conservation with conductances for Na^+ and K^+ ions through the membrane that depend on the voltage $V(t)$. This reflects the change in permeability to these ions of proteins that transect the membrane and change their conformation as a function of the voltage across the membrane. The specific forms of the voltage dependent conductivities in this HH model are taken from textbook descriptions based on the 1940s and 1950s work of Hodgkin, Huxley, Katz, and many others. The reversal potentials are determined by the competition of diffusion associated with ion concentration differences within and without the cell and transport of charged ions by the electric field associated with the difference in voltage across the membrane. The Nernst equation which determines these reversal potentials is directly from statistical physics. The cell responds to external currents as a driving force by its cross membrane voltage $V(t)$ rising if the current causes depolarization of the cell, or the voltage decreases when the cell becomes more polarized. The rise in voltage triggers an instability in the phase space of the HH model associated with a sudden influx of Na^+ ions which is then counterbalanced by a flow of K^+ ions out of the cells as the voltage rises to order $+ 50$ mV. All this takes place on the order of 5 ms and is seen as a 'spike' in the voltage time series.

The model is governed by the following four first order differential equations:

$$C\frac{dV(t)}{dt} = I_{inj}(t) + g_{Na}m(t)^3 h(t)(E_{Na} - V(t))$$

$$+ g_K n(t)^4 (E_K - V(t)) + g_L(E_L - V(t))$$

$$\frac{da(t)}{dt} = \frac{a_0(V(t)) - a(t)}{\tau_a(V(t))} \qquad a(t) = \{m(t), h(t), n(t)\}$$

$$a_0(V) = \frac{1}{2} + \frac{1}{2}\tanh\left(\frac{V - V_a}{\Delta V_a}\right)$$

$$\tau_a(V) = \tau_{a0} + \tau_{a1}\left(1 - \tanh^2\left(\frac{V - V_a}{\Delta V_a}\right)\right). \tag{5.19}$$

In these equations the g_{ion}'s are maximum conductances for the ion channels, the E_{ion} are reversal potentials for those ion channels, $I_{inj}(t)$ is the external stimulating current injected into the neuron. This current is selected by the experimenter and has no independent dynamics.

The gating variables $a(t)$ are taken to satisfy first order kinetic equations and range between zero and unity. The overall strength of an ion channel is set by the maximal conductances, and this represents the number of individual ion channels. These are phenomenological choices.

The quantities $a_0(V)$ and $\tau_a(V)$ are the voltage dependent activation function and the voltage dependent time constant of the gating variable $a(t)$. The external forcing of the cell $I_{inj}(t)$ is known to us. This current is selected by the user. In mainstream neurobiology it is quite often taken to be a sequence of square pulses, but here we take a selection of such pulses along with segments of a chaotic waveform designed to probe the dynamical range of neuron voltage response. The injected current should also have significant low frequency content as the cell's membrane acts as a capacitor with a resistance with characteristics of a low pass filter. High frequency components of $I_{inj}(t)$ are effectively filtered out by the cell membrane and not delivered to the dynamics of the cell's electrical response in action potentials.

Only the voltage across the cell membrane is measurable in real neurobiological experiments. Successful data assimilation, in effect, 'measures' the gating variable time series as well as the unknown parameters. We present only noisy time series of $V(t)$ to the model; these are our $\mathbf{y}(t)$ in the notation we have used for the general discussion above.

In our parametrization of the cell dynamics there are 19 fixed parameters and three unobserved state variables $a(t) = \{m(t), h(t), n(t)\}$ to be determined. All $a(t)$ lie between zero and one. Over the observation window $[t_0, t_{final}]$ we must present enough information via the measurements $\mathbf{y}(\tau_k)$ to accurately estimate all of these quantities at t_{final}. We then can use the values at this final time of observations as initial conditions for prediction of the response of the neuron model for $t > t_{final}$.

The parameters used to generate data are listed in Table 5.1. The waveform of the injected current is chosen to be a combination of step functions and segments of a chaotic time series. This current is displayed in the bottom panel of Fig. 5.5. A standard adaptive fourth order Runge-Kutta solver (Press et al. (2007)) is used to produce the data using time steps of $\Delta t = 0.025$ms, and white Gaussian noise with an RMS level of 1mV is added to the $V(t)$ time series to represent the noise accompanying the measurements in laboratory biological experiments. This voltage time course is in the top panel of Fig. 5.5.

The previous example of the Lorenz96 $D = 5$ model contains only quadratic nonlinear terms in their differential equations. The difficulties of state and parameter estimation result from their chaotic trajectories. The NaKL model in the selected parameter region is not chaotic (Guckenheimer and Oliva (2002)), and the challenge of data assimilation comes from the richer nonlinearity in the dynamics of the gating variables and the sensitivity of the model behavior to changes in parameter values.

In the numerical optimization used to find the saddle paths for any model one must specify search bounds for each parameter and each state variable. The goal is to find appropriate choices for these values that constrain the model states and parameters to biologically acceptable regions. The bounds for the voltage $V(t)$ are taken as -150mV and $+70$mV based upon our solutions to the equations. The gating variables are bounded between 0 and 1, since they represent the probability whether ion channels are open or closed.

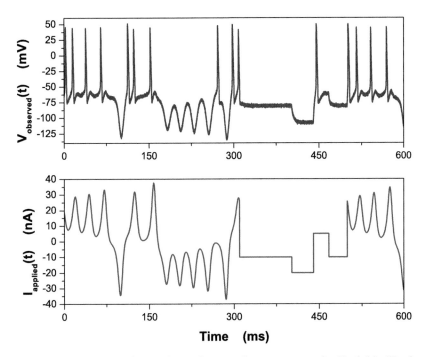

Figure 5.5 **Top Panel** Observed membrane voltage response for Hodgkin-Huxley NaKL model. **Bottom Panel** Stimulating current $I_{inj}(t)$ injected into the NaKL model: Eq. (5.19).

The optimization is performed with IPOPT using an interior-point method (Wachter and Biegler (2006)). We found the interior-point method both more stable and substantially faster than the L-BFGS-B method (Zhu et al. (1997)).

In this twin experiment only a noisy voltage $V(t)$ is 'measured' and presented to the model; so $L = 1$. As the dynamical range of voltage is a hundred times larger than that of the gating variables, we first calculated the action levels with $R_m^{(V)} = 1$, $R_{f0}^{(V)} = 10^{-3}$, $R_{f0}^{(m)} = 10$, $R_{f0}^{(h)} = 10$, $R_{f0}^{(n)} = 10$, and $\alpha = 3/2$. The largest β was taken as 50. Also we decreased α from 2 to $3/2$ so that the pace of increasing resolution in model state space is slower than in our earlier examples as we increment changes in R_f. This allows us to stay well within the basin of attraction of the lowest action level.

The top panel of Fig. 5.6 displays the action level plot with the configuration above. When $25 \leq \beta \leq 39$, there are several different levels, and they reveal the expected action level after $\beta \geq 40$. The action level plot suggests that the voltage measurement alone is sufficient to determine the three unobserved state variables as well as the 19 parameters. The estimates of these parameters are displayed in Table 5.1.

We can select the values of R_{f0}'s for each state variable according to our knowledge about the amplitudes and time scales of the state variables by looking

Table 5.1 *Known and estimated parameters for the NaKL model. We require bounds to be specified for the nonlinear search algorithm, and these are shown. For the state variables these are taken from the data, and for the gating variables the bounds are [0,1]. Maximal conductances $\{g_{Na}, g_K, \dots\}$ must be positive.*

Parameters	Known	Estimated	Lower Bound	Upper Bound
g_{Na}	120.0	108.4	50.0	200.0
E_{Na}	50.0	49.98	0.0	100.0
g_K	20.0	21.11	5.0	40.0
E_K	−77.0	−77.09	−100.0	−50.0
g_L	0.3	0.3028	0.1	1.0
E_L	−54.0	−54.05	−60.0	−50.0
C	0.8	0.81	0.5	1.5
V_m	−40.0	−40.24	−60.0	−30.0
ΔV_m	0.0667	0.0669	0.01	0.1
τ_{m0}	0.1	0.0949	0.05	0.25
τ_{m1}	0.4	0.4120	0.1	1.0
V_h	−60.0	−59.43	−70.0	−40.0
ΔV_h	−0.0667	−0.0702	−0.1	−0.01
τ_{h0}	1.0	1.0321	0.1	5.0
τ_{h1}	7.0	7.76	1.0	15.0
V_n	−55.0	−54.52	−70.0	−40.0
ΔV_n	0.0333	0.0328	0.01	0.1
τ_{n0}	1.0	1.06	0.1	5.0
τ_{n1}	5.0	4.97	2.0	12.0

at the time series of solutions of the model. The time constants of gating variables characterize their response to the change of voltage. The sodium activation variable $m(t)$ is the fastest, with a time constant of several hundreds of microseconds, which is a little bit slower than $V(t)$. $h(t)$ and $n(t)$ are much slower, having a time constant of a few milliseconds. We set the ratio of $R_{f0}^{(m)}/R_{f0}^{(V)} = 5 \times 10^4$ and raised the ratio of $R_{f0}^{(n)}/R_{f0}^{(m)}$ and $R_{f0}^{(h)}/R_{f0}^{(m)}$ from 1 to 10 to compensate for the effects induced by different time constants. With $R_m = 1$, $R_{f0}^{(V)} = 10^{-3}$, $R_{f0}^{(m)} = 50$, $R_{f0}^{(h)} = 500$, $R_{f0}^{(n)} = 500$, the action level plot in the bottom panel of Fig. 5.6 shows this configuration of R_{f0} can effectively enforce that most saddle paths stay near the expected lowest action level. The detailed action levels can depend on the choice of R_{f0}. In the NaKL example, when the ratios of $R_{f0}^{(n)}/R_{f0}^{(m)}$ and $R_{f0}^{(h)}/R_{f0}^{(m)}$ are as large as 100, we often observed another action level with a value close to the lowest one. This can also depend on the specific choice of the N_O initial paths with which we start the annealing.

A lesson we learn from this example is that we should take both the amplitude and the time scale of the state variables into consideration when selecting the scale of R_f values. The proper configurations of R_f values enlarge the probability to

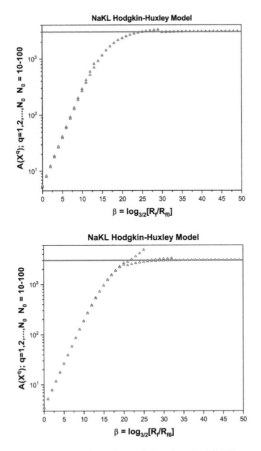

Figure 5.6 Action levels as a function of R_f for the NaKL model with only the noisy membrane voltage $V(t)$ measured and presented to the model. **Top Panel**: we selected $R_{f0}^{(V)} = 10^{-3}$, $R_{f0}^{(m)} = 10$, $R_{f0}^{(h)} = 10$, $R_{f0}^{(n)} = 10$ and $\alpha = 3/2$; **Bottom Panel**: $R_m = 1$, $R_{f0}^{(V)} = 10^{-3}$, $R_{f0}^{(m)} = 10$, $R_{f0}^{(h)} = 1000$, $R_{f0}^{(n)} = 1000$, and $\alpha = 3/2$.

have the candidate paths converge to the expected action level, and also accelerate the convergence rate to the optimal paths.

5.4 Qualitative Commentary about Precision Annealing

Many readers will be familiar with the idea of *simulated annealing* (Kirkpatrick et al. (1983); Press et al. (2007)) and recognize precision annealing as a relative. In simulated annealing the entire action is scaled by a 'temperature,' more or less equivalent to $1/R_f$, and whatever calculations are desired are repeated as we move from large 'temperature' (small R_f) to small 'temperature.'

In data assimilation we have an additional term which results from measurements carrying information about the desired state space locations of the model at

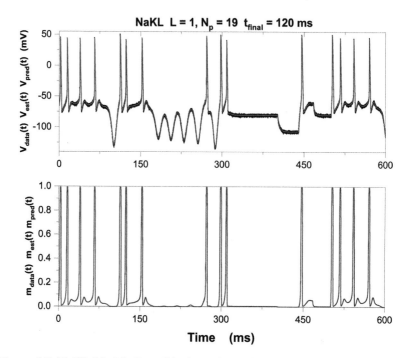

Figure 5.7 NaKL Model: Data (black), estimated (red) and predicted (blue) state variables $V(t)$, $m(t)$ for the NaKL model when only the noisy membrane voltage $V(t)$ is measured and presented to the model. Refer back to Fig. 5.5 and Table 5.1.

specific times. When, in the standard model, $R_f \to 0$, the state space information from observations imposes a harmonic oscillator 'potential' on the motions in the model state space along the directions of the measurements $y_l(\tau_k)$. It leaves the motion untouched and uninformed in the remaining unobserved state space directions in which it acts as a free particle. This potential constrains the motion in the observed directions, and constrains more and more directions as L increases.

Moving about as a free particle gives the motion very little structure in all unobserved directions and transfers information about state space locations but not about model parameters θ. As we increase R_f, more or less similar to the mass of the particle with coordinates $\mathbf{x}(t)$, the particle motion is constrained to move along orbits in path space dictated by the dynamics and thus slowly providing structure to the paths along which the particle motion $\mathbf{x}(t)$ may proceed.

When β in our annealing schedule reaches β_{max} (or equivalently the maximum value of R_f), we have imposed enough structure on the paths to represent the information content of the measurements within the model constraints, and no further dynamical constraint is required.

While this commentary has some flaws (for example, at $R_f = 0$ there is no 'kinetic energy' for the particles), it helps to recall that the action is

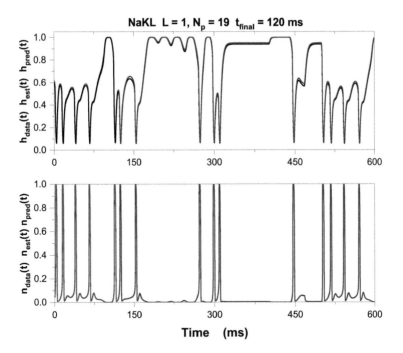

Figure 5.8 NaKL model: Data (black), estimated (red) and predicted (blue) state variables $h(t)$, $n(t)$ when only the noisy membrane voltage $V(t)$ is measured and presented to the model. Refer back to Fig. 5.5 and Table 5.1.

$A(\mathbf{X}) = -\log[P(\mathbf{X}|\mathbf{Y})]$, and its average over the ensemble of paths is an entropy or self-information of the system. In Chapter 8 of this monograph we will find this commentary hopefully useful in interpreting what machine learning algorithms are accomplishing through training.

6

Discrete Time Integration in Data Assimilation Variational Principles: Lagrangian and Hamiltonian Formulations

In this chapter we explore the *symplectic symmetries* (Gelfand and Fomin (1963); Goldstein et al. (2002); Arnol'd (1989)) that are a general part of the variational principles used in data assimilation and elsewhere. These may also be familiar from Classical Mechanics. In geosciences the name **4DVar** has long been used for the variational formulation of DA, but the issue of whether it respects the symplectic symmetry within the calculation to improve prediction accuracy is typically overlooked.

This was discussed in Section 3.3.3 partly as a preparation for the results here. There we focused on the symplectic symmetry in Hamiltonian coordinates $\{\mathbf{x}, \mathbf{p}\}$, but since these are reached from the coordinates in which we typically encounter mechanics, namely, D-dimensional position $\mathbf{x}(t)$ and D-dimensional velocity $d\mathbf{x}(t)/dt = \dot{\mathbf{x}}(t)$, via a Legendre transformation, we expect these formulations of mechanics, or variational problems in general (Gelfand and Fomin (1963); Kot (2014); Liberzon (2012); Goldstein et al. (2002)), we choose to start there.

There are many books, one reaching back to 1904 (Whittaker and McCrae (1988)), and others more recent (Gelfand and Fomin (1963); Goldstein et al. (2002); Arnol'd (1989); Kot (2014); Liberzon (2012); Leok and Zhang (2011)), presenting the material we introduce. Any exploitation of the stability and numerical robustness gained respecting the symplectic symmetry in numerically integrating the dynamical equations of motion, the Euler-Lagrange equations, is, to our knowledge, more or less absent in most of these references.

There seems little, if any, recognition of these symplectic symmetries in data assimilation (but see Kadakia et al. (2017)) and machine learning. We take up this topic again in Chapter 8.

6.0.1 Canonical and Not

Variational principles have a natural symplectic structure among the state variables $\mathbf{x}(t)$ and $\dot{\mathbf{x}}(t) = d\mathbf{x}/dt$ (Whittaker and McCrae (1988); Gelfand and Fomin (1963)).

The implications of this structure in both Lagrangian coordinates $\{\mathbf{x}(t), \dot{\mathbf{x}}(t)\}$ and Hamiltonian canonical coordinates $\{\mathbf{x}(t), \mathbf{p}(t)\}$ are explored in this chapter. We illustrate this using a detailed numerical examination of questions raised by assuring that the symplectic structure of orbits in the chaotic Lorenz 1996 model with D = 10 dimensions. There are a number of differences associated with the coordinate choice, and canonical coordinates can sometimes improve the quality of the predictions.

The contemporary interest in the symplectic structure of dynamical systems associated with variational principles arose when accurate long time predictions and stability of orbits were required in a variety of applications, and mainstream integration methods failed. This is a selection of fields where that was seen: accelerator Physics, plasma Physics, and stability of planetary orbits as investigated circa 1980–90. (Sanz-Serna (1992); Channell and Scovel (1990); Ruth (1983); Wisdom and Holman (1991)). When issues in those fields arose in solving specific classical mechanics problems, especially focusing on energy conservation, restoration of symplectic symmetry of the discrete time formulation of the dynamical rules solved the issues.

Guided by the equivalence of the Lagrangian and the Hamiltonian formulation in classical mechanics and in data assimilation – and as discussed in Chapter 8 – machine learning, we explore the data assimilation action in the Hamiltonian description where the symplectic symmetry manifests itself most transparently, and in Lagrangian coordinates where the symplectic symmetry manifests itself in a bit more subtle manner.

The *canonical* coordinates $\{\mathbf{x}(t), \mathbf{p}(t)\}$ in which the Hamiltonian $H(\mathbf{x}, \mathbf{p})$ operates, exhibit the symplectic structure, responsible for the preservation of quantities called symplectic bilinear forms along Hamiltonian flows. These invariants assume a simple form as phase space areas. While it is often unmentioned in physics texts, the equivalent Lagrangian description in $\{\mathbf{x}(t), \dot{\mathbf{x}}(t)\}$ coordinates also exhibits symplectic structure, but it is a bit more complex, as the symplectic bilinear form is not a directed area in $\{\mathbf{x}(t), \dot{\mathbf{x}}(t)\}$ (Marsden and West (2001)).

The usual discussion of symplectic symmetry in classical mechanics texts presents it as a feature in flows in *continuous* time. Numerical solutions of Hamiltonian problems must reside in *discrete* time. Formulating or imposing symplectic symmetry on solutions carried out in discrete time requires a bit of care.

It is well established that solving the Euler-Lagrange equations in a way that preserves the symplectic invariants provides significantly improved numerical stability and accuracy (Chyba et al. (2009); Wisdom and Holman (1991)).

Application of the methods described here to variational data assimilation had not appeared in the literature before Kadakia et al. (2017). We note, however, that a paper by Sanz-Serna (2016) has shown there are benefits to preserving symplectic structure in adjoint methods for uncertainty quantification.

We discuss the implications of symplectic structure in both Lagrangian and Hamiltonian descriptions to variational data assimilation.

- First, while the invariance of symplectic bilinear forms is guaranteed along continuous flows, maintaining this invariance along discretized paths in time requires more attention. We discuss how to preserve this invariance by building on integrators applicable directly to Hamilton's equations, as well as those naturally suited to discrete actions, known as variational integrators (Marsden and West (2001)).
- Second, as variational data assimilation is a two-point boundary value problem for expected values for functions of $\mathbf{X}$, we study the form of these boundary conditions and how to most appropriately enforce them.

There are a variety of differences between the Hamiltonian and Lagrangian approaches. For instance, the boundary conditions are explicitly enforced in the Hamiltonian approach, but require constrained optimization. The Lagrangian version, on the other hand, is unconstrained, and the boundary conditions are satisfied implicitly by the discrete time Euler-Lagrange equations.

6.0.2 The Laplace Approximation

We focus on the approximation of the expected value integral Eq. (3.1) using Laplace's method. This is a variational calculation seeking extremum paths $\mathbf{X}^q$, $q = 0, 1, 2, \ldots$ of the action. As we are interested specifically in minima, we seek solutions to the following:

$$\frac{\partial A(\mathbf{X})}{\partial \mathbf{X}}\bigg|_{\mathbf{X}=\mathbf{X}^q} = 0 \quad \text{and} \quad \frac{\partial^2 A(\mathbf{X})}{\partial \mathbf{X}^2}\bigg|_{\mathbf{X}=\mathbf{X}^q} > 0. \tag{6.1}$$

The second inequality indicates that eigenvalues of the Hessian matrix must be positive.

When the dynamical equations

$$\frac{dx_a(t)}{dt} = F_a(\mathbf{x}(t), \boldsymbol{\theta}, t) \tag{6.2}$$

or

$$x_a(n + 1) = f_a(\mathbf{x}(n), \boldsymbol{\theta}) \tag{6.3}$$

are nonlinear, there may be multiple solutions to Eq. (6.1). These solutions often have distinct values of the action $A(\mathbf{X})$, which provide exponentially separated approximations to the expected value. We call $\mathbf{X}^1$ the path(s) that give the smallest value of the action $A(\mathbf{X}^1)$

In nonlinear problems the search for $\mathbf{X}^1$, the path yielding the smallest action, requires some care. We discussed this requirement using an annealing method in Chapter 5 introduced in Ye et al. (2015a,b) to locate the extremum paths with the smallest action. We return to this annealing method, as well as an adaptation of it for canonical coordinates, shortly.

6.0.3 The Standard Model Action

Our discussion proceeds in the framework of the *standard model* action. The model error is taken to be distributed as a Gaussian with mean zero and precision R_f. The measurement error is taken to be distributed as a Gaussian with mean zero and precision R_m.

Together, these assumptions lead to the *Standard Model Action* in discrete time:

$$A(\mathbf{X}) = \sum_{n=0}^{N} \sum_{l=1}^{L} \frac{R_m(l)}{2} \left(x_l(n) - y_l(n) \right)^2 \tag{6.4}$$

$$+ \sum_{n=0}^{N-1} \sum_{a=1}^{D} \frac{R_f(a)}{2} \left(x_a(n+1) - f_a(\mathbf{x}(n), \mathbf{x}(n+1), \boldsymbol{\theta}) \right)^2.$$

Estimating state variables (measured and unmeasured) as well as time independent model parameters $\boldsymbol{\theta}$ is the core challenge in data assimilation through approximating Eq. (3.1), as $\mathbf{F}(\mathbf{x}(t), \boldsymbol{\theta}, t)$ is nonlinear in the $\mathbf{x}(t)$.

6.1 Symplecticity in Variational Data Assimilation

We turn to the role of symplectic structure in variational data assimilation. These remarks are intended to be general, in order to raise awareness of results from continuous time classical and discrete time classical mechanics, and their connection with data assimilation.

6.1.1 Continuous Time

We find it simpler to address the ideas first in continuous time before treating the discrete time case, which has some subtle but important differences.

The Hamiltonian Formulation

In continuous time, the action has the form

$$A(\mathbf{X}) = \int_{t_0}^{t_{final}} dt \, L(\mathbf{x}(t), \dot{\mathbf{x}}(t), t), \tag{6.5}$$

where $L(\mathbf{x}(t), \dot{\mathbf{x}}(t), t)$ is the Lagrangian. As discussed in Ye et al. (2015a,b), the Lagrangian for the standard model action Eq. (6.4) is given by

$$L(\mathbf{x}(t), \dot{\mathbf{x}}(t), t) = \sum_{l=1}^{L} \frac{R_m(t)}{2} \Big(x_l(t) - y_l(t) \Big)^2$$
$$+ \sum_{a=1}^{D} \frac{R_f(a)}{2} \Big(\dot{x}_a(t) - F_a(\mathbf{x}(t)) \Big)^2 + \frac{\nabla \cdot \mathbf{F}(\mathbf{x}(t), \boldsymbol{\theta})}{2}, \quad (6.6)$$

and, when the measurement error term is absent, it is also called the Onsager-Machlup functional (Onsager and Machlup (1953)). As we have noted, the integral for $E[G(\mathbf{X})|\mathbf{Y}]$ is not Gaussian when $\mathbf{F}(\mathbf{x})$ is nonlinear. This expression provides a straightforward derivation of the Hamiltonian function via a Legendre transformation from generalized coordinates $\{\mathbf{x}, \dot{\mathbf{x}}\}$ to phase space canonical coordinates $\{\mathbf{x}, \mathbf{p}\}$. The canonical momenta (Goldstein et al. (2002); Arnol'd (1989)) $\mathbf{p}(t)$ are defined as

$$p_a(t) = \frac{\partial L(\mathbf{x}(t), \dot{\mathbf{x}}(t), t)}{\partial \dot{x}_a(t)} = R_f(a) \big(\dot{x}_a(t) - F_a(\mathbf{x}(t), \boldsymbol{\theta}) \big), \quad (6.7)$$

for the standard model action.

The Hamiltonian $H(\mathbf{x}(t), \mathbf{p}(t), t)$ is determined via the Legendre transform $\{\mathbf{x}, \dot{\mathbf{x}}\} \rightarrow \{\mathbf{x}, \mathbf{p}\}$ to be

$$H(\mathbf{x}(t), \mathbf{p}(t), t) = \sum_{a=1}^{D} \dot{x}_a(\mathbf{x}(t), \mathbf{p}(t), t) \cdot p_a(t)$$
$$- L(\mathbf{x}(t), \dot{\mathbf{x}}(\mathbf{x}(t), \mathbf{p}(t), t), t). \quad (6.8)$$

For the standard model action, this change of coordinates $\{\mathbf{x}(t), \dot{\mathbf{x}}(t)\} \rightarrow \{\mathbf{x}(t), \mathbf{p}(t)\}$ is possible since the precision $\mathbf{R}_f$ is positive definite. We easily find

$$\frac{dx_a(t)}{dt} = F_a(\mathbf{x}(t), \boldsymbol{\theta}, t) + \frac{p_a(t)}{R_f(a)}, \quad (6.9)$$

and the Hamiltonian for Eq. (6.6) becomes

$$H(\mathbf{x}(t), \mathbf{p}(t), t) = \sum_{a=1}^{D} \left(\frac{p_a(t)^2}{2R_f(a)} + p_a(t) F_a(\mathbf{x}(t), \boldsymbol{\theta}) \right)$$
$$- \sum_{l=1}^{L} \frac{R_m(l)}{2} \Big(x_l(t) - y_l(t) \Big)^2 - \frac{\nabla \cdot \mathbf{F}(\mathbf{x}(t), \boldsymbol{\theta})}{2} \quad (6.10)$$

or, for brevity,

$$H(\mathbf{x}(t), \mathbf{p}(t), t) = \sum_{a=1}^{D} \left(\frac{p_a(t)^2}{2R_f(a)} + p_a(t) F_a(\mathbf{x}(t), \boldsymbol{\theta}) \right)$$
$$-\chi(\mathbf{x}(t) - \mathbf{y}(t)) - \frac{\nabla \cdot \mathbf{F}(\mathbf{x}(t), \boldsymbol{\theta})}{2}. \tag{6.11}$$

In canonical coordinates, the equations of motion are given by Hamilton's equations (Whittaker and McCrae (1988); Goldstein et al. (2002))

$$\frac{dx_a(t)}{dt} = \frac{\partial H(\mathbf{x}(t), \mathbf{p}(t), t)}{\partial p_a(t)}$$
$$-\frac{dp_a(t)}{dt} = \frac{\partial H(\mathbf{x}(t), \mathbf{p}(t), t)}{\partial x_a(t)}, \tag{6.12}$$

which for the standard model action are

$$\frac{dx_a(t)}{dt} = F_a(\mathbf{x}(t), \boldsymbol{\theta}) + \frac{p_a(t)}{R_f(a)}$$

$$-\frac{dp_a(t)}{dt} = \sum_{b=1}^{D} p_b(t) \frac{\partial F_b(\mathbf{x}(t))}{\partial x_a(t)}$$

$$-\sum_{l=1}^{L} R_m(l) \left(x_l(t) - y_l(t)\right) \delta_{al} - \frac{1}{2} \frac{\partial \nabla \cdot \mathbf{F}(\mathbf{x}(t), \boldsymbol{\theta})}{\partial x_a(t)}$$

$$-\frac{dp_a(t)}{dt} = \sum_{b=1}^{D} p_b(t) \frac{\partial F_b(\mathbf{x}(t))}{\partial x_a(t)}$$

$$-\frac{\partial \left[\chi(\mathbf{x}(t) - \mathbf{y}(t)) + \frac{\nabla \cdot \mathbf{F}(\mathbf{x}(t), \boldsymbol{\theta})}{2} \right]}{\partial x_a(t)}. \tag{6.13}$$

Solutions to Eq. (6.12) and Eq. (6.13) generate a Hamiltonian flow in continuous time, along which the following bilinear form is invariant (Arnol'd (1989)):

$$\omega_{H(\mathbf{x}, \mathbf{p})} = dx_a \wedge dp_a. \tag{6.14}$$

This is the symplectic bilinear form (or 2-form) Arnol'd (1989) in canonical coordinates and describes a directed area of phase space. In addition to this, $H(\mathbf{x}(t), \mathbf{p}(t))$ is conserved along orbits, phase space volume $d^D\mathbf{x}d^D\mathbf{p}$, and the Poincaré invariants are conserved as well (Goldstein et al. (2002); Arnol'd (1989)).

As we noted earlier, these conserved quantities are associated with the Poisson brackets among functions $A(\mathbf{x}, \mathbf{p})$ and $B(\mathbf{x}, \mathbf{p})$ in $\{\mathbf{x}, \mathbf{p}\}$ space:

$$\left\{ A(\mathbf{x}, \mathbf{p}), B(\mathbf{x}, \mathbf{p}) \right\} \Big|_{\mathbf{x}, \mathbf{p}} = \sum_{a=1}^{D} \left[\frac{\partial A(\mathbf{x}, \mathbf{p})}{\partial x_a} \frac{\partial B(\mathbf{x}, \mathbf{p})}{\partial p_a} - \frac{\partial A(\mathbf{x}, \mathbf{p})}{\partial p_a} \frac{\partial B(\mathbf{x}, \mathbf{p})}{\partial x_a} \right]. \tag{6.15}$$

The equations of motion for $A(\mathbf{x}, \mathbf{p}))$ are

$$\frac{dA(\mathbf{x}(t), \mathbf{p}(t))}{dt} = \left\{ A(\mathbf{x}(t), \mathbf{p}(t)), H(\mathbf{x}(t), \mathbf{p}(t)) \right\} \Bigg|_{\mathbf{x}(t), \mathbf{p}(t)}, \tag{6.16}$$

so $dH(\mathbf{x}(t), \mathbf{p}(t))/dt = 0$. The conservation of phase space volume follows directly from Eq. (6.12).

The Lagrangian Formulation

We now discuss the somewhat lesser known appearance of symplectic structure in the Lagrangian description. As usual in the calculus of variations (Gelfand and Fomin (1963)), the necessary conditions for an extremum of the action in Eq. (6.5), derived via the variational principle, are the Euler-Lagrange equations in $\{\mathbf{x}, \dot{\mathbf{x}}\}$ space

$$\frac{d}{dt} \left(\frac{\partial L(\mathbf{x}(t), \dot{\mathbf{x}}(t), t)}{\partial \dot{x}_a(t)} \right) = \frac{\partial L(\mathbf{x}(t), \dot{\mathbf{x}}(t), t)}{\partial x_a(t)}, \tag{6.17}$$

along with the boundary conditions that one must satisfy, arising from an integration by parts:

$$\sum_{a=1}^{D} \delta x_a(t_0) p_a(t_0) = 0 \quad \sum_{a=1}^{D} \delta x_a(t_{final}) p_a(t_{final}) = 0. \tag{6.18}$$

For the standard model action, the Euler-Lagrange equations are

$$\frac{d^2 x_a(t)}{dt^2} = \sum_{b=1}^{D} \Omega_{ab}(\mathbf{x}(t)) \frac{dx_b(t)}{dt}$$

$$+ \frac{\partial [\mathbf{F}(\mathbf{x}, \boldsymbol{\theta})^2 / 2 + \chi(\mathbf{x}(t) - \mathbf{y}(t)) + \frac{\nabla \cdot \mathbf{F}(\mathbf{x}(t), \boldsymbol{\theta})}{2}]}{\partial x_a(t)} \tag{6.19}$$

and

$$\Omega_{ab}(\mathbf{x}(t)) = \frac{\partial F_a(\mathbf{x}(t), \boldsymbol{\theta})}{\partial x_b} - \frac{\partial F_b(\mathbf{x}(t), \boldsymbol{\theta})}{\partial x_a} \tag{6.20}$$

is a skew-symmetric matrix.

While the Hamiltonian 2-form Eq. (6.14) is not invariant along integral curves of the Euler-Lagrange equations, there does exist an invariant symplectic bilinear form in generalized coordinates $\{\mathbf{x}, \dot{\mathbf{x}}\}$ (Wendlandt and Marsden (1997)):

$$\omega_L(\mathbf{x}, \dot{\mathbf{x}}) = \frac{\partial^2 L(\mathbf{x}(t), \dot{\mathbf{x}}(t), t)}{\partial x_a \partial \dot{x}_b} dx_a \wedge dx_b$$

$$+ \frac{\partial^2 L(\mathbf{x}(t), \dot{\mathbf{x}}(t), t)}{\partial \dot{x}_a \partial \dot{x}_b} d\dot{x}_a \wedge dx_b. \tag{6.21}$$

This implies that Lagrangian flows are symplectic. Since Lagrangian flows are equivalent to Hamiltonian flows, one might suspect both have preserved symplectic forms; these are they.

Noting that

$$dp_a(\mathbf{x}, \dot{\mathbf{x}}) = \frac{\partial p_a(\mathbf{x}, \dot{\mathbf{x}})}{\partial x_b} dx_b + \frac{\partial p_a(\mathbf{x}, \dot{\mathbf{x}})}{\partial \dot{x}_b} d\dot{x}_b, \qquad (6.22)$$

the preserved Lagrangian symplectic form is equivalent to the usual canonical preserved symplectic form in Hamiltonian dynamics, Eq. (6.14).

6.1.2 Discrete Time

Each of these continuous time formulations need to be translated into discrete time when numerical evaluations are required. Fortunately this has been analyzed quite extensively (Marsden and West (2001); Hairer et al. (2006)). Unexpected effects such as artificial energy dissipation and poor long-term behavior were the first indicators that difficulties arise when Hamilton's equations are integrated numerically with a choice of integration algorithms that does not respect the underlying symplectic symmetry, Eq. (3.34). The fundamental issue is that symplectic structure is not always preserved when discretizing Hamilton's equations. This can be avoided by choosing judicious integration schemes for the discrete time implementation of the Euler-Lagrange equations of motion, producing a class of integrators called *symplectic integrators*. Some common integration schemes, such as the Euler methods and the fourth order Runge-Kutta method, are not symplectic, while others, such as the modified midpoint rule are; see p. 171 of Hairer et al. (2006).

A note is useful here. We do not require the term $\nabla \cdot \mathbf{F}(\mathbf{x}, \boldsymbol{\theta})/2$ that arises from the singular limit $\Delta t \to 0$, as Δt is always nonzero.

Numerical integrators that preserve symplectic bilinear forms arise quite naturally from discretizing the action directly, rather than from the equations of motion (Marsden and West (2001)). Such variational integrators are well-suited to the standard model action of statistical data assimilation, which is already formulated in discrete time. This is the viewpoint we adopt in the Lagrangian description. As we will see, the different discretization methods for the Hamiltonian and Lagrangian formulations are well-suited to each, and provide two consistent yet distinct approaches to the data assimilation problem.

Discrete Time Lagrangian Formulation

Marsden and West (2001) show that given a discretized action of the form

$$A(\mathbf{X}) = \sum_{n=0}^{N-1} L_d(\mathbf{x}(n), \mathbf{x}(n+1)), \qquad (6.23)$$

the discrete time variational principle

$$\frac{\partial A(\mathbf{X})}{\partial \mathbf{X}} = 0, \tag{6.24}$$

produces $N - 1$ Euler-Lagrange equations for $n = 1, \ldots, N - 1$

$$\frac{\partial L_d(\mathbf{x}(n-1), \mathbf{x}(n))}{\partial \mathbf{x}(n)} + \frac{\partial L_d(\mathbf{x}(n), \mathbf{x}(n+1))}{\partial \mathbf{x}(n)} = 0 \tag{6.25}$$

as well as boundary conditions at t_0 and $t_N = t_{final}$

$$\frac{\partial L_d(\mathbf{x}(0), \mathbf{x}(1))}{\partial \mathbf{x}(0)} = 0 \quad \frac{\partial L_d(\mathbf{x}(N-1), \mathbf{x}(N))}{\partial \mathbf{x}(N)} = 0. \tag{6.26}$$

Along maps satisfied by the discrete time Euler-Lagrange equations Eq. (6.25), the following discrete symplectic bilinear form is preserved:

$$\omega_{L_d}(\mathbf{x}(n), \mathbf{x}(n+1)) = \sum_{i,j=1}^{D} \frac{\partial^2 L_d}{\partial x_i(n)\, \partial x_j(n+1)} dx_i(n) \wedge dx_j(n+1).$$

The discrete time variational principle Eq. (6.24) produces simultaneously:

- a symplectic mapping,
- the proper boundary conditions, and
- the discrete Euler-Lagrange equations.

These are a natural result of setting the gradient of the action (Eq. (6.23)) equal to zero, so any local or global minimum of the action will satisfy these conditions.

It is a key requirement that the action sum depends on *both* $\mathbf{x}(n)$ and $\mathbf{x}(n+1)$, and that dependence treats them as independent variables in the variation of the action $\delta A(\mathbf{X})$. As explained earlier this introduces into a discrete time formulation the 2D degrees of freedom inherent in continuous time Lagrangian dynamics: $\{\mathbf{x}(t), \dot{\mathbf{x}}(t)\}$. This is as it should be, as the act of discretizing the label called time should not change the number of degrees of freedom in the problem. The discrete time standard model action (Eq. (6.4)), has this form, and the corresponding discrete Lagrangian is

$$L_d(\mathbf{x}(n), \mathbf{x}(n+1)) = \sum_{l=1}^{L} \frac{R_m(n)}{2} \left(x_l(n) - y_l(n) \right)^2$$

$$+ \sum_{a=1}^{D} \frac{R_f(a)}{2} \left(x_a(n+1) - f_a(\mathbf{x}(n), \mathbf{x}(n+1), \boldsymbol{\theta}) \right)^2. \tag{6.27}$$

Discrete Time Hamiltonian Formulation

The discrete time variational principle can also be applied to the Hamiltonian description to produce variational integrators for Hamiltonian systems. In the context of discrete mechanics, it was introduced by Marsden and West (2001) and later extended by Lall and West (2006) to give discrete analogs of Hamilton's equations. All this (including extensions to discrete Hamilton-Jacobi theory, which we have not considered) can be formulated nicely using generating functions for canonical transformations (Goldstein et al. (2002)). See Leok and Zhang (2011) for a particularly nice overview.

A discrete time version of Hamilton's equations can be derived from a discrete time Legendre transform. In contrast to the continuous time construction, there are two choices of discretization. One can define respectively the *right and left Legendre transforms*

$$p_a^+(n+1) = +\frac{\partial L_d(\mathbf{x}(n), \mathbf{x}(n+1))}{\partial x_a(n+1)}$$

$$p_a^-(n) = -\frac{\partial L_d(\mathbf{x}(n), \mathbf{x}(n+1))}{\partial x_a(n)}, \tag{6.28}$$

which have corresponding right and left discrete forms of the Hamiltonian

$$H_d^+(\mathbf{x}(n), \mathbf{p}(n+1)) = \sum_{a=1}^{D} p_a(n+1) \, x_a(n+1)$$
$$- L_d(\mathbf{x}(n), \mathbf{x}(n+1))$$

$$H_d^-(\mathbf{x}(n+1), \mathbf{p}(n)) = -\sum_{a=1}^{D} p_a(n) \, x_a(n)$$
$$- L_d(\mathbf{x}(n), \mathbf{x}(n+1)). \tag{6.29}$$

The respective quantities $\mathbf{x}(n+1)$ and $\mathbf{x}(n)$ are given *implicitly* by the right and left discrete Hamilton's equations

$$x_a(n+1) = \frac{\partial H_d^+(\mathbf{x}(n), \mathbf{p}(n+1))}{\partial p_a(n+1)}$$

$$p_a(n) = \frac{\partial H_d^+(\mathbf{x}(n), \mathbf{p}(n+1))}{\partial x_a(n)} \tag{6.30}$$

$$x_a(n) = -\frac{\partial H_d^-(\mathbf{x}(n+1), \mathbf{p}(n))}{\partial p_a(n)}$$

$$p_a(n+1) = -\frac{\partial H_d^-(\mathbf{x}(n+1), \mathbf{p}(n))}{\partial x_a(n+1)}. \tag{6.31}$$

Both versions generate maps that preserve the discrete time bilinear symplectic forms

$$\omega_{H_d^+}(\mathbf{x}(n), \mathbf{p}(n+1)) = dx_a(n) \wedge dp_a(n+1) \tag{6.32}$$

$$\omega_{H_d^-}(\mathbf{x}(n+1), \mathbf{p}(n)) = dx_a(n+1) \wedge dp_a(n), \tag{6.33}$$

which are in fact equivalent: $\omega_{H_d^+}(\mathbf{x}(n), \mathbf{p}(n+1)) = \omega_{H_d^-}(\mathbf{x}(n+1), \mathbf{p}(n))$.

So, the right and left formulations are equivalent, in the sense that they both produce mappings that preserve symplectic structure, along with the discrete Hamilton's equations

$$x_a(n+1) = f_a(\mathbf{x}(n)) + \frac{p_a(n+1)}{R_f(a)}$$

$$p_a(n) = \sum_{b=1}^{D} \frac{\partial f_b(\mathbf{x}(n))}{\partial x_a} p_b(n+1)$$

$$+ \sum_{l=1}^{L} R_m(l)(x_l(n) - y_l(n)) \delta_{al}. \tag{6.34}$$

These equations, which are quite similar to their continuous time counterparts Eq. (6.13), are known from optimal control (Gelfand and Fomin (1963)) and estimation theory, and describe explicit schemes in which the states are mapped forwards in time $\mathbf{x}(n) \to \mathbf{x}(n+1)$ and the canonical momenta (or co-states) are mapped backwards in time $\mathbf{p}(n+1) \to \mathbf{p}(n)$. On the other hand, when the forward mapping is not explicit, the discrete Hamiltonian cannot be written down in explicit form, because we cannot generally invert the right discrete Legendre transform, solving for $\mathbf{x}(n+1)$ in terms of $\mathbf{x}(n)$ and $\mathbf{p}(n+1)$.

This construction provides a powerful framework for deriving variational integrators for discrete time Hamiltonian systems. Although it is formulated in terms of a discrete Lagrangian, this is not required. As shown in Leok and Zhang (2011), it may also be obtained as a direct variational discretization of continuous Hamiltonian mechanics. The variational framework is particularly well-suited to our form of the standard model action (Eq. (6.4)), and we have adopted it in our previous algorithm (Ye et al. (2015a,b)), which is cast as a discrete Lagrangian.

We now discuss how to reformulate this algorithm in the Hamiltonian description, but will not treat it as a variational integrator. We take a simpler approach, by directly discretizing Hamilton's equations in continuous time, and then comparing the difference between using symplectic and non-symplectic integrators. This reformulation offers some interesting trade-offs.

6.2 A Symplectic Annealing Method

In the Lagrangian case, the boundary conditions and symplectic structure are coupled together in the search for minimizing paths. By contrast, in the Hamiltonian formulation the transition to a $2D$-dimensional phase space $\{\mathbf{x}, \mathbf{p}\}$ decouples these constraints from the overall search and provides more explicit control over how they are enforced.

With this in mind, we describe a method for finding minimizing paths of the action when constrained by annealing its symplectic structure. The annealing is intended to help mitigate some of the issues due to the existence of multiple local minima of the non-convex action, which is especially prevalent when the estimation window is long, the dynamics are chaotic and the observations are sparse. The intuition for this stems from our previous approach, which performs a similar type of annealing in the discrete Lagrangian description (Ye et al. (2015a,b)). We briefly recap the idea first before exploring the method in detail.

6.2.1 Lagrangian Approach

The method introduced in Ye et al. (2015a,b) directly minimizes the discretized standard model action (Eq. (6.4)) through nonlinear optimization, in which the states $\mathbf{X} = \{\mathbf{x}(n)\}$; $n = 0, 1, \ldots, N$ at different times are treated as independent variables. This approach can be considered a variational formulation derived from a discrete Lagrangian, and is part of the family of multiple shooting and collocation methods for solving optimal control problems (Betts (2010)).

The main idea, and explained at some length in Chapter 5, is to treat the magnitude of the model error term R_f as an annealing hyperparameter, whose value starts off at small R_{f_0} and is then gradually increased over sequential iterations of the optimization routine with each subsequent iteration initialized from the previous solution. We choose $R_f = R_{f_0} \alpha^\beta$ so at each iteration in $\beta = 0, 1, 2, \ldots$ the magnitude of R_f increases by a factor α; $\alpha > 1$. Initially, since R_f is small, the estimated path closely matches the observations and is highly degenerate in the unmeasured variables. As its value increases, the model error term is given more weight and this degeneracy is broken. This procedure has been seen to be more efficient than direct optimization of the nonlinear action at large R_f in locating the lowest minimum of the action, as well as in quantifying the contribution of minimum paths to the conditional expectation integral Eq. (3.1). The process is carried out to a high value of R_f and may be consistent with the assumed confidence in the model dynamics. For the numerical experiments considered here, the final value of R_f will be rather high, since our data will be generated from a deterministic model that is assumed known. In theory this corresponds to the limit $R_f \to \infty$, but finite precision prevents this limit from being realized computationally. Since R_f has no 'true' fixed value in this context, there is no problem treating it as an annealing parameter for the estimation process. When this process is successful, the action level will stabilize at some finite value of R_f, and the action becomes essentially independent of R_f.

As mentioned above, in the Lagrangian description, the boundary conditions and symplectic structure of the solution are imposed implicitly by the discrete time variational principle (Eq. (6.24)), and therefore cannot be treated separately by the optimization algorithm. This is not so in the Hamiltonian formulation, which will allow us to utilize the model dynamics in a slightly different way. The main trade-off with this approach is that it requires adding constraints to the optimization procedure.

6.2.2 Hamiltonian Approach in Canonical Coordinates

The analogous program in the Hamiltonian description is derived not from the variational approach discussed in Section 6.1.2, but rather from a direct discretization of the continuous time Hamilton's equations Eq. (6.13). We discuss our choices of discretization in a moment, but first focus on the core ideas.

Hamilton's equations reformulate the necessary conditions for a stationary path, described by the Euler-Lagrange equations in the Lagrangian context, in $2D$-dimensional phase space $\{\mathbf{x}, \mathbf{p}\}$. These equations may be solved (in discretized form) as a nonlinear system of equations, and represent constraints on the space of possible solutions. In this formulation, the boundary conditions may be applied explicitly since the canonical momenta are treated as independent variables, along with the states. We are not, however, interested in finding *any* minimizing path, but rather focus on those paths that produce the lowest minima. With this in mind, we express the continuous time action Eq. (6.4) in canonical coordinates by substituting for the canonical momenta Eq. (6.7), and then write the integral as a discrete time sum

$$A(\mathbf{X}, \mathbf{P}) = \sum_{n=0}^{N} \left(\sum_{a=1}^{D} \frac{p_a(n)^2}{2 R_f(a)} + p_a(n) f_a(\mathbf{x}(n), \boldsymbol{\theta}) \right.$$
$$\left. + \sum_{l=1}^{L} \frac{R_m(l)}{2} \left(x_l(n) - y_l(n) \right)^2 \right). \tag{6.35}$$

This expression for the action is then used as the objective function for *constrained optimization*, conditioned on a discretized form of Hamilton's equations (6.13) as constraints. This form is purely quadratic (assuming a linear measurement function); the nonlinearity in the model has been transferred to the constraints.

Embedding the original action Eq. (6.4) in $2D$-dimensional phase space provides more explicit control over how the constraints of the underlying variational principle are satisfied. For instance, the boundary conditions may be enforced directly as a constraint on the initial and final canonical momenta $\mathbf{p}(0) = \mathbf{p}(N) = 0$, since they are now independent variables in the optimization. Also, relaxing these constraints to inequalities allows us to anneal them explicitly, and provides the same functionality of the Lagrangian approach – by slowly introducing the dynamical constraints of the model – in a more direct way.

The Hamiltonian annealing method may be summarized as follows. Fixing R_f and the constraints on the initial and the terminal canonical momenta to some small tolerance $|p_a(0)| \leq g_p$ and $|p_a(N)| \leq g_p$, we slowly anneal constraints given by a discretization $\phi(\cdot)$ of Hamilton's equations.

We discuss two different discretizations of the Hamiltonian equations of motion:

- one is the trapezoidal rule. This is not a symplectic integration rule!
- the other is the modified midpoint rule. This **is** a symplectic integration rule.

6.3 Three Integration Methods

We have already described the Lagrangian Method. It is the variational principle with annealing in R_f.

In canonical coordinates the Hamiltonian is

$$H(\mathbf{x}(t), \mathbf{p}(t), t) = \sum_{a=1}^{D} \left[\frac{p_a(t) p_a(t)}{2 R_f(a)} + p_a(t) F_a(\mathbf{x}(t), \boldsymbol{\theta}, t) \right]$$

$$-\chi(\mathbf{x}(t) - \mathbf{y}(t))$$

and

$$\chi(\mathbf{x}(t) - \mathbf{y}(t)) = \sum_{l=1}^{L} \frac{R_m(l)}{2} (\mathbf{x}_l(t) - \mathbf{y}_l(t))^2. \tag{6.36}$$

The **midpoint integration–symplectic** integrator (Hairer et al. (2006)). It uses annealing on the precision with which the boundary conditions are met: $|\mathbf{p}(0)| \leq g_H = \frac{g_{H0}}{\alpha^\beta}$ and $|\mathbf{p}(N)| \leq g_H = \frac{g_{H0}}{\alpha^\beta}$.

$$\mathbf{x}(n+1) = \mathbf{x}(n) + \Delta t \frac{\partial H(\mathbf{x}(t), \mathbf{p}(t), t)}{\partial \mathbf{p}} \bigg|_{\mathbf{x} = \mathbf{x}(n+1/2), \mathbf{p} = \mathbf{p}(n+1/2)}$$

and

$$\mathbf{p}(n+1) = \mathbf{p}(n) - \Delta t \frac{\partial H(\mathbf{x}(t), \mathbf{x}(t), t)}{\partial \mathbf{x}} \bigg|_{\mathbf{x} = \mathbf{x}(n+1/2), \mathbf{p} = \mathbf{p}(n+1/2)} \tag{6.37}$$

In these expressions for any variable: $x(n + 1/2) = \frac{1}{2}[x(n) + x(n + 1)]$.

The **trapedoidal integration–not symplectic** method follows. It uses annealing on the precision with which the boundary conditions are met:

$$\mathbf{x}(n+1) = \mathbf{x}(n) + \frac{\Delta t}{2} \left\{ \frac{\partial H(\mathbf{x}(t), \mathbf{p}(t), t)}{\partial \mathbf{p}} \bigg|_{\mathbf{x}=\mathbf{x}(n)), \mathbf{p}=\mathbf{p}(n)} + \right.$$

$$\left. \frac{\partial H(\mathbf{x}(t), \mathbf{p}(t), t)}{\partial \mathbf{p}} \bigg|_{\mathbf{x}=\mathbf{x}(n+1)), \mathbf{p}=\mathbf{p}(n+1)} \right\}$$

and

$$\mathbf{p}(n+1) = \mathbf{p}(n) - \frac{\Delta t}{2} \left\{ \frac{\partial H(\mathbf{x}(t), \mathbf{p}(t), t)}{\partial \mathbf{x}} \bigg|_{\mathbf{x}=\mathbf{x}(n)), \mathbf{p}=\mathbf{p}(n)} + \right.$$

$$\left. \frac{\partial H(\mathbf{x}(t), \mathbf{p}(t), t)}{\partial \mathbf{x}} \bigg|_{\mathbf{x}=\mathbf{x}(n+1)), \mathbf{p}=\mathbf{p}(n+1)} \right\}. \tag{6.38}$$

This also uses annealing on the precision with which the boundary conditions are met: $|\mathbf{p}(0)| \leq g_H = \frac{g_{H0}}{\alpha^\beta}$ and $|\mathbf{p}(N)| \leq g_H = \frac{g_{H0}}{\alpha^\beta}$.

The most notable difference between the two methods is that the Hamiltonian approach involves constrained optimization while the Lagrangian is unconstrained. As a result, the Lagrangian version requires considerably less computation since it does not have to deal with extra complexity of nonlinear constraints. The numerical experiments we present show that the Hamiltonian version is more accurate in certain circumstances. The trade-off between computational speed and accuracy should be evaluated on a case-by-case basis.

In the Lagrangian description, the gradual increase of R_f deforms the action manifold in path space $A(\mathbf{X})$ from one that is highly quadratic in the measured components and nearly flat in the unmeasured components, to one rendered highly non-convex by the nonlinearity of the vector field $\mathbf{F}(\mathbf{x}(t), \boldsymbol{\theta})$. The incremental way in which this is done attempts to track the local minima of the action systematically, even if for large R_f they occupy tiny, deep corners in the action manifold. At small R_f, due to the approximate flatness of the action in the unmeasured directions, the stationary paths are highly degenerate, and this degeneracy is slowly lifted with increasing R_f using our schedule protocol for R_f. Meanwhile, as discussed above, the symplecticity of every path of minimum action is guaranteed at each step in $\beta = \log_\alpha[R_f/R_{f0}]$.

In the Hamiltonian formulation the path is embedded in a $2D$-dimensional phase space of canonical coordinates. Part of the usefulness of this description is the calculation can be carried out in a way that preserves symplectic structure. However, the use of equality constraints makes it difficult to slowly enforce the model dynamics, since solutions to Hamilton's equations coincide with the action minima in the first place.

We use the constraints as an annealing device to slowly impose the model dynamics, gradually removing degeneracies related to the partial observability ($L \leq D$) of the system. At low β, stationary paths are degenerate in the measured components, and $\mathbf{p}(n) \approx 0$. Because the momenta appear quadratically, and because their restriction through the equations of motion is not yet imposed, this degeneracy is present for arbitrary R_f. It is then gradually lifted through the annealing process by slowly enforcing the model dynamics through the application of Hamilton's equations as constraints.

Unlike the Lagrangian method where R_f changes at each iteration, the action functional does not change throughout this annealing process. Instead, the feasible region is deformed, which has the effect of gradually enforcing the symplecticity of the Hamiltonian mappings. This effectively decouples the variational calculation from the boundary conditions and the symplecticity of the flow, so they may be enforced directly as constraints on the optimization.

We turn to some of the details associated with the implementation of these methods. The numerical optimization was performed using IPOPT (Wachter and Biegler (2006)). All derivatives were coded manually. Since the Lagrangian version is unconstrained, the algorithm essentially performs Newton's method, though we did impose static constraints on the state variables to keep them within the dynamical range of the model.

For the Hamiltonian approaches, R_f was fixed at 10^6 throughout the annealing, equal to the final value of R_f for the Lagrangian method, so model uncertainty is the same for all methods at the end of the annealing procedure. Further details are listed in Table 6.1.

6.4 Numerical Twin Experiments

We evaluate the performance of the three variants of these annealing methods with a set of numerical twin experiments on the Lorenz96 model (Lorenz (2006)) in $D = 10$ dimensions. The dynamical equations for the model are given by

$$\frac{dx_a}{dt} = x_{a-1}(x_{a+1} - x_{a-2}) - x_a + \theta. \tag{6.39}$$

The state variables are periodic $x_{D+1} = x_1$, and the forcing parameter $\theta = 8.17$ is chosen so the trajectories are chaotic (Lorenz (2006); Lorenz and Emanuel (1998)).

We examine the effectiveness of these methods as a function of the number of observations, L, by choosing $P = 100$ random initial conditions $\{\mathbf{x}_1^*(0), \ldots, \mathbf{x}_{100}^*(0)\}$. Each is integrated forward using an explicit fourth order Runge-Kutta scheme with a time step of $\Delta t = 0.01$ to produce P sets of time series to be used as the data. Choosing the length of the estimation window to be $t_{final} = 4$ 'observations' are made of the first $L = 1, 2, 3, 4,$ or 5 components of

Table 6.1 *The three annealing protocols. Limits: $R_f = 10^6$, $g_p = 10^{-4}$*

Method	Annealing	α	β	Initial
$L_{\text{variational}}$	$R_f = R_{f_0}\alpha^\beta$	$10^{1/4}$	$0\ldots 40$	$R_{f0} = 10^{-4}$
$H_{\text{non-symplectic}}$, $H_{\text{symplectic}}$	$g_H = g_{H0}/\alpha^\beta$	$10^{1/6}$	0.36	$g_{H0} = 10^2$

these trajectories. These are then corrupted with additive Gaussian noise of mean zero and unit variance.

These noisy trajectories are sampled at either $\Delta t = 0.01$ or $\Delta t = 0.05$. For $\Delta t = 0.01$ the observations contain $N = 401$ data points, and $N = 81$ data points when we select $\Delta t = 0.05$. The result is a set of $P = 100$ distinct noisy data sets $\{\mathbf{Y}_1, \ldots, \mathbf{Y}_{100}\}$ sampled from different locations around the attractor. To correspond to an actual field or laboratory observation, only the first L components of each of dataset were used in the data assimilation procedure.

These 'data' (this is a twin experiment), along with the known dynamical equations, are provided as input, via the vector field $\mathbf{F}(\mathbf{x}(t), \boldsymbol{\theta}, t)$ to the three annealing methods. Each produces estimations and predictions of the full path $\mathbf{X}$, containing both the L measured and the $D - L$ unmeasured components, as well as the time independent parameters $\boldsymbol{\theta}$. Since the data were generated from the model equations themselves, these simulations, despite the fact that the action (Eq. (6.4)) contains model errors quantified by R_f. This is not inconsistent, however, as our annealing methods take R_f from quite small to quite large $\sim 10^6$. In Ye et al. (2015a,b) this was found to be a more effective way to approach the limit $R_f \to \infty$ than imposing strong model constraints throughout. This is especially true when the assimilation window is long, the model is chaotic and the observations are sparse (Quinn (2010)).

For each data set $\mathbf{Y}_p$; $p = 1, 2, \ldots, 100$, we initialize the optimization from a set of $q = 1, \ldots, N_0 = 100$ distinct initial path estimates $\mathbf{X}_q^{(0)}$ sampled uniformly across the dynamical range of the attractor, roughly $[-10, 10]$.

The annealing is then performed for each of these $N_I = 100$ initial selections, producing a set of estimated paths $\mathbf{X}_{q,p}^{(\beta)}$. The same set of initial trajectories are used for each data set, so $\mathbf{X}_q^{(0)} \equiv \mathbf{X}_{q,p}^{(0)}$. The final estimates are denoted $\mathbf{X}_{q,p}^{(\beta_{\text{final}})}$.

For each unique data path $\mathbf{Y}_p$ we produce a progression of path estimates $\mathbf{X}_{q,p}^{(\beta)}$ for each initial condition q. The $N_I = 100$ initial conditions provide some idea of how reliable the algorithm is, and the $P = 100$ distinct datasets give statistical information across the entire attractor. The former allows us to address questions such as: given no prior information about the system state, how often is an algorithm able to find the lowest action level? The latter provides confidence that our conclusions about a particular method are not sensitive to local properties such as model instability or system observability.

6.4.1 Action Level Plots

We first utilize these data by plotting their corresponding action values as a function of β. These "action level plots" allow us to visualize the global structure of the action and understand its dependence on various parameters of the estimation, such as the number of observations.

Fig. 6.1 (**Upper Panel**) displays the action levels from the Lagrangian algorithm as a function of $\beta = \log_\alpha[R_f/R_{f0}]$ for $L = 1$. The different colors on the plot indicate different selections for q among the 100 initial guesses, but otherwise the colors hold no significance. The degeneracy of the action levels for many

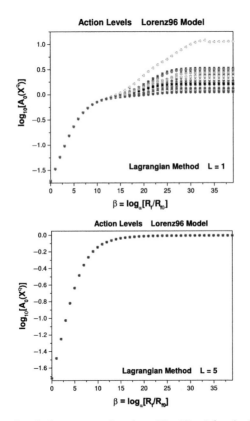

Figure 6.1 Action level plots versus $\beta = \log_\alpha[R_f/R_{f0}]$ for the Lorenz96 model, $D = 10$, for **Upper Panel** $L = 1$ and for **Lower Panel** $L = 5$. $R_{f0} = 10^{-4}$ and $\alpha = 10^{1/4}$. These results use the Lagrangian variation of the standard action. The results for $L = 1$ indicate that one measurement at each observation time is not enough to produce well separated action levels. This is an indication of many local minima of the action. At $L = 5$ these local minima, resulting from instabilities on the synchronization manifold of the data and the model outputs, are resolved and a single accurate minimum of the action remains.

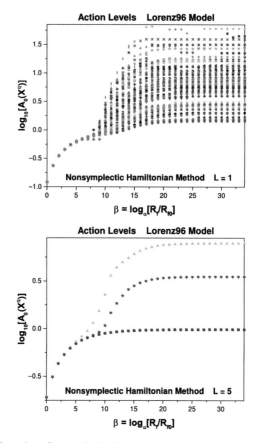

Figure 6.2 Action plots for a calculation as in Fig. 6.1. We use the *non-symplectic Hamiltonian* annealing protocol of the standard action; **Upper Panel** $L = 1$ measured variables and **Lower Panel** $L = 5$ measured variables.

paths is observed for small β, and as β increases the levels split, revealing the complex structure of the action surface. Many of the initial paths go to the same minimum.

Fig. 6.1 (**Lower Panel**) shows the results of the same calculation for $L = 5$ observed state variables. As the number of measurements increases, the information provided by the data can stabilize the search for local minima, rendering the action surface either nearly convex or largely attractive in the basin of the smallest minimum. These plots are from Ye et al. (2015a,b).

Fig. 6.2 (**Upper Panels**) (**Nonsymplectic Hamiltonian Method**) and Fig. 6.3 (**Symplectic Hamiltonian Method**) show corresponding results for the equality constraint annealing, using the nonsymplectic and symplectic methods, respectively. Fig. 6.2 and Fig. 6.3 are for $L = 1$, and reveal that the action has substantially richer structure when represented in canonical coordinates. When

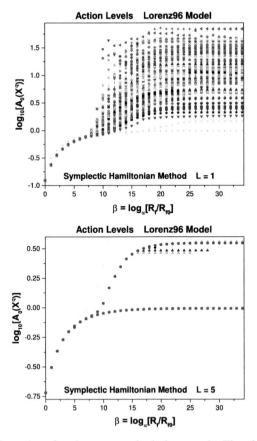

Figure 6.3 Action plots for the same calculations as in Fig. 6.1 and Fig. 6.2, using the symplectic version of the Hamiltonian annealing protocol; **Upper Panel** $L = 1$ measured component and **Lower Panel** $L = 5$ measured components.

action levels are close together, Laplace's method requires summing over a rather large number of paths to accurately approximate the conditional expected value.

In Fig. 6.2 and Fig. 6.3 (**Lower Panels**) the number of measured variables is increased to $L = 5$, and the structure of minima in the action appears much simpler. The large separation between the lowest action minimum and the next lowest action minimum indicates that the lowest minimum path alone provides an accurate approximation. Therefore, despite the presence of higher action levels that do not appear in the Lagrangian annealing, both the Hamiltonian methods should produce accurate predictions when $L = 5$.

In comparing the difference between the two Hamiltonian methods, it may come as a surprise that a symplectic and a nonsymplectic method produce similar results. We explain this by noting that the trapezoidal rule is a conjugate symplectic integrator, which, although not symplectic, is approximately symplectic (Hairer et al. (2006)). The difference in annealing protocol between the Lagrangian and Hamiltonian methods may also have more impact on the qualitative structure of the action levels than the choice of integration scheme.

6.4.2 Illustrative Estimates and Predictions

We now look at the quality of estimates and predictions of both measured and unmeasured variables for the three integration methods. The estimates and predictions for a particular dataset $\mathbf{Y}_p$ are calculated using the single estimated path $\mathbf{X}_p^{\min}$ that gives the lowest action level of each of the $N_0 = 100$ initial conditions. The predicted trajectories are then computed by integrating forward the estimated state at the end of the estimation window as initial conditions.

Fig. 6.4 shows estimates for $0 \leq t \leq 4$ to the left of the vertical line and predictions for $t > 4$ to the right, both for $L = 1$. In the upper panel is the observed variable $x_2(t)$ and in the lower panel is $x_8(t)$, which is unobserved. Neither the estimates nor the predictions are particularly accurate, in agreement with our conclusions from the action levels plots for $L = 1$. Fig. 6.5 displays the same results but for $L = 5$. For both the observed and unobserved variable, we see a substantial improvement in both the estimates and the predictions. Since the quality of the predictions relies on estimating the full model state at the end of the observation window, we infer that for $L = 1$ the unmeasured variables are not accurately estimated, but for $L = 5$ they are well estimated.

Estimation and Prediction Errors

The plots shown so far give a rough idea of the qualitative differences between the three variational protocols as the number of measured components is varied, with regard to the action surface structure and prediction quality. We now give a more quantitative comparison in terms of the attractor-averaged, point-wise, mean-squared error

$$E(t_n) = \frac{1}{DP} \sum_{p=1}^{P} \sum_{a=1}^{D} \left(x_{p,a}^*(t_n) - x_{p,a}^{\min}(t_n) \right)^2, \tag{6.40}$$

where $x_{p,a}^*(t_n)$ are the known components of the 'true' data paths $\mathbf{X}_p^*$, and $x_{p,a}^{\min}(t_n)$ are the components of the estimate $\mathbf{X}_p^{\min}$ with the lowest action level. This statistic provides a global estimate for how the algorithm performs within the observation window, at its boundaries, and during prediction.

In Fig. 6.6 (**Left Panel**) we display the pointwise estimation error (Eq. (6.40)) for $L = 1$ with $\Delta t = 0.05$.

The error metric Eq. 6.40 for estimates and predictions are shown in Fig. 6.6 and 6.6, respectively. It appears the Lagrangian approach has the advantage here,

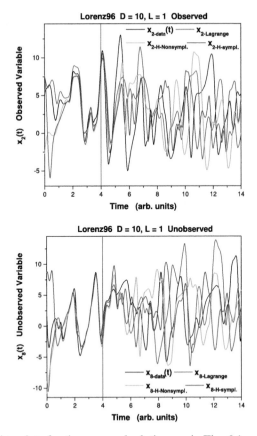

Figure 6.4 Action plots for the same calculations as in Fig. 6.1 and Fig. 6.2, using the three versions of our annealing protocols; **Upper Panel** $L = 1$ an observed component $x_2(t)$ and **Lower Panel** $L = 5$ an unobserved component $x_8(t)$. We display for each of $x_2(t)$, $x_8(t)$: the data (in black), the estimated values for $t \leq 4$, and the predicted value for $t > t_{final} = 4.0$.

though these errors are rather large, compared with the observational noise and subsequent results for larger L.

Fig. 6.7 repeats these calculations for $L = 4$. The lowest action levels for the $P = 100$ different locations on the attractor have now become somewhat less cluttered, indicating all the algorithms had more success in finding the global minimum. This is further supported by the fact that the estimation and prediction error have also decreased by about an order of magnitude. The Lagrangian method appears to give slightly better estimates, but the predictions are all about equivalent.

All methods show a saturation of the prediction error around $t = 7$. A similar saturation occurs in Figure 6.6, but it is not as apparent because the predictions are only shown from $t = 4$ to $t = 6$. This arises due to the chaos in the Lorenz96 system.

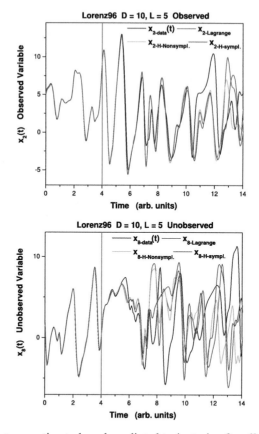

Figure 6.5 The true, estimated, and predicted trajectories for all three annealing protocols. These are for the same calculation as Fig. 6.4, with $L = 5$ observed components of the noisy data. We display for an observed variable $x_2(t)$ and for an unobserved variable $x_8(t)$, the data (in black), the estimated value for $t \leq 4$, and the predicted value for $t > t_{final} = 4.0$.

Small errors in the estimate are exponentially magnified until all trajectories are dispersed around the attractor and the error saturates at a value commensurate with the range of the dynamical variables.

In Fig. 6.8 we repeat these calculations again, but for $L = 5$. The estimation error is roughly the same as in Fig. 6.7. Although the non-symplectic Hamiltonian method has begun to match the performance of the Lagrangian variational method, all methods give similar quality estimates at $t = 4$, and there is little discernible difference among the predictions.

Figures 6.9, 6.10, and 6.11 repeat these same calculations with a smaller time step $\Delta t = 0.01$. Most of the comments, as for the three previous figures, hold for the $L = 1$ case, shown in Fig. 6.9. No method is able to produce accurate

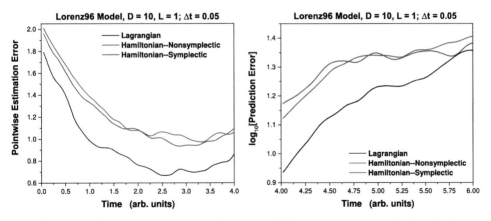

Figure 6.6 For the calculation with $\Delta t = 0.05$, and $L = 4$ we display **Left Panel** the pointwise average estimation error (Eq. 6.40) as a function of t_n in the estimation window $0 \leq t_n \leq 4$, and **Right Panel** the pointwise average prediction error for $t > 4$, for each of the three different annealing methods. In it is seen that for any given q, there are many distinct lowest minimum values. The Lagrangian method seems to yield the smallest estimation and prediction errors; however, since only $L = 1$ out of 10 variables is measured, all estimations are not particularly accurate.

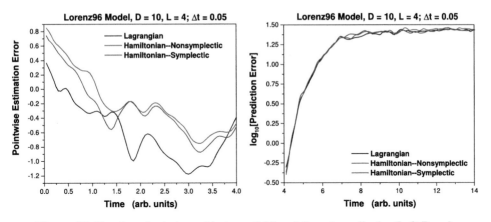

Figure 6.7 For the calculation with $\Delta t = 0.05$ and $L = 4$ we display **Left Panel**, the pointwise average estimation error (Eq. (6.40) as a function of t_n in the estimation window $0 \leq t_n \leq 4$, and in the **Right Panel** the pointwise average prediction error for $t > 4$, for each of the three different annealing methods.

estimates or predictions, and the Lagrangian approach has a slight advantage over the Hamiltonian methods.

As with $\Delta t = 0.05$, the situation changes for $L = 4$ and $L = 5$, and a distinct advantage of the Hamiltonian method has now emerged. For instance, with $L = 4$ in Fig. 6.10 the Lagrangian method is unable to find the lowest action

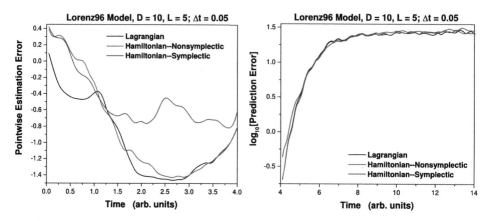

Figure 6.8 The calculations shown here are identical to those in Fig. 6.6 and Fig. 6.7, except now with $L = 5$ measured variables. The estimation error is again somewhat lower for the Lagrangian method in regions of the observation window, but the error at $t_{final} = 4$ is nearly identical in all cases, so the prediction errors are indistinguishable among the three cases.

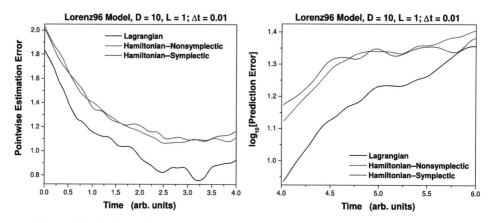

Figure 6.9 In these plots we repeat the calculations of Fig. 6.6, except we reduce the timestep of observations from $\Delta t = 0.05$ to $\Delta t = 0.01$. There is only one, $L = 1$, measured component. Similar to the plots in Fig. 6.6, the prediction and estimation errors are quite high, with the Lagrangian method showing a small advantage over the Hamiltonian methods.

level in several instances. The Hamiltonian nonsymplectic method on the other hand consistently returns lowest minima near unity for all but a few locations, and the Hamiltonian symplectic method performs even better. As a result, the average estimation and prediction errors shown in Fig. 6.10 are noticeably lower for the symplectic Hamiltonian approach. In the estimation window, these errors are reduced by nearly two orders of magnitude.

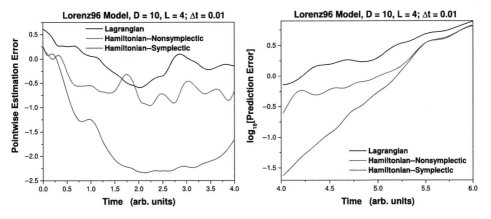

Figure 6.10 At $L = 4$, $\Delta t = 0.01$ the nonsymplectic Hamiltonian method exhibits the properties in Fig. 6.9 far less regularly, and the Hamiltonian symplectic annealing method, barely at all. The result is average estimations errors appreciably lower for the symplectic Hamiltonian method (up to two orders of magnitude in the estimation window), with correspondingly reduced prediction errors as well.

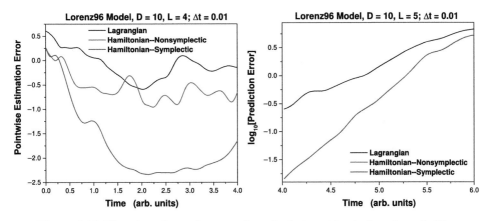

Figure 6.11 The plots shown here are for calculations identical to those in Fig. 6.9 and Fig. 6.10, with $\Delta t = 0.01$, except now we have increased the number of measured variables to $L = 5$. A similar phenomenon as for the $L = 4$ case occurs. Accordingly, estimation and prediction errors for the Hamiltonian actions are appreciably lower for both Hamiltonian methods.

Fig. 6.11 shows a similar situation occurs for $L = 5$. The two Hamiltonian methods produce identical estimations and predictions, which are quite a bit lower than the Lagrangian case.

We now examine the details of the estimation produced from the 13th initial path on the attractor for $L = 5$. An action level plot for this case is shown in Fig. 6.12. For all the $N_0 = 100$ initial paths the Lagrangian method only identifies a single

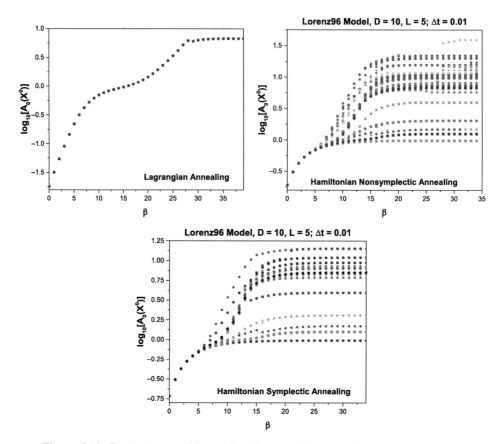

Figure 6.12 For the Lorenz96 model with $D = 10$, $\Delta t = 0.01$, and $L = 5$ we display the full action plots for the 13th attractor locations, whose lowest minima is plotted in Fig. 6.11. In the **Upper Left Panel** is the full action plot for the Lagrangian annealing protocol. $\beta = \log_\alpha[R_f/R_{f0}]$. In the **Upper Right Panel** we show that of the Hamiltonian nonsymplectic annealing method, and in the **Lower Panel** we show the action plot of the Hamiltonian symplectic annealing protocol. $\beta = \log_\alpha[g_{H0}/g_H]$. While the Lagrangian annealing in R_f finds only a single minimum value at high β, this value is far outside the range of the consistent normalized lowest value of 1.0. Conversely, the Hamiltonian annealing protocol finds many local minima but consistently returns a lowest value near 1.0.

action level, significantly above the expected lowest value near unity. Meanwhile, the Hamiltonian methods find many local minima, including this expected lowest value, suggesting that in certain circumstances, a more disperse set of action levels is useful for identifying the global minimum.

6.5 Summary of Symplectic Annealing Methods

We have investigated different methods for implementing the variational method to estimate the conditional expected value (Eq. (3.1)) of functions $G(\mathbf{X})$ along the path of a model state through observation and prediction windows. The variational principle is familiar from many sources, Gelfand and Yaglom (1960); Kot (2014); Liberzon (2012) among others, and it is well established that finding the minimum of the action, the log-likelihood of the conditional probability distribution $A_0(\mathbf{X}) = -\log[P(\mathbf{X}|\mathbf{Y})]$, is a two-point boundary value problem. In the meteorological literature, this is known as 4DVar (Evensen (2009); Bennett (1992); Chua and Bennett (2001)).

For the usual formulation in $\{\mathbf{x}(t), \dot{\mathbf{x}}(t)\}$ space, the variational principle provides the Euler-Lagrange equation with boundary conditions that require the canonical momentum to vanish at the endpoints of the observation window. We recalled previous results that show how a local minimum of the action is guaranteed to satisfy both the Euler-Lagrange equations and the boundary conditions, as well as preserve a symplectic bilinear form. In discrete time, these considerations produce variational integrators, which have been remarkably successful in producing stable numerical calculations over long time windows. Moreover, they appear to be naturally suited to variational data assimilation, which has the inherent structure of a discrete Lagrangian.

We also recalled an alternative description, in canonical coordinates $\{\mathbf{x}, \mathbf{p}\}$, which preserves symplectic structure, but decouples the solution of Hamilton's equations from the minimization of the action and the enforcement of the boundary conditions. Motivated by this observation, we investigated an alternative method, precision annealing, described in Chapter 5.

This Hamiltonian formulation comes with a number of inherent trade-offs. Given a linear measurement function, the action becomes quadratic in $\{\mathbf{x}, \mathbf{p}\}$ space, the boundary conditions $\mathbf{p}(0) = \mathbf{p}(N) = 0$ may be directly constrained, and symplecticity explicitly enforced with a suitable integration scheme. The problem then involves nonlinear constraints, so it requires more computational effort, but these constraints can be annealed directly, without having to modify the model error term characterized by R_f at each step.

We performed a set of numerical experiments to investigate the difference between the Lagrangian formulation and two distinct Hamiltonian versions: one with a symplectic integrator and one without. For the chaotic Lorenz96 model with $D = 10$, we found that Hamiltonian methods produce better estimates and predictions when the time step is small $\Delta t \approx 0.01$ and the system is moderately observable ($L = 4$ or $L = 5$), and Lagrangian methods appear to work better when Δt is larger. Moreover, the Hamiltonian action appears to have a more disperse set of local minima, which is evidently helpful in some cases for finding the global minima.

The fact that the two annealing methods return different solutions to the optimization problem is not surprising, given the distinct ways in which they lift the degeneracy associated with partial system observability. For the Lagrangian annealing, the gradual deformation from a near-quadratic to non-convex action surface may confine the search to a particular region of state space in which there exists a local minimum with a strong basin of attraction. This deformation process may play a role in collecting the admissible solutions to a small neighborhood of the high-dimensional action manifold early in the annealing process, *before* the finer features of the action materialize at higher R_f.

Why does the nonsymplectic version work just as well as the symplectic one? We mentioned that the nonsymplectic trapezoidal integrator happens to be a conjugate symplectic integrator. As discussed in Hairer et al. (2006) and Hairer and Zbinden (2012), conjugate symplectic methods, while not precisely preserving the symplectic symmetry of the differential equations, do produce high accuracy, long-term integrations of these problems, and may be sufficient for the problem at hand.

Why does the Hamiltonian approach produce more disperse action level plots? There is an inversion that occurs when moving to canonical coordinates. The model error term involves $\mathbf{R}_f^{-1}$, instead of $\mathbf{R}^f$ as in the Lagrangian description. When $\mathbf{R}_f$ is small, the momenta should be small so the measurement error term dominates the action. As $\mathbf{R}_f$ increases, the momenta are allowed to take on larger values, but must find their way back to a neighborhood of the origin, as that is a necessary condition for a global minimum in the deterministic problem (Liberzon (2012)).

7

Monte Carlo Methods

The main objective in data assimilation has been to evaluate expected values of functions $\langle G(\mathbf{X}) \rangle$ on paths in path (state variable, parameter) space

$$
\begin{aligned}
\langle G(\mathbf{X}) \rangle &= \int d\mathbf{X}\, P(\mathbf{X}|\mathbf{Y}) G(\mathbf{X}) \\
&= \frac{\int d\mathbf{X}\, G(\mathbf{X}) \exp[-A(\mathbf{X})]}{\int d\mathbf{X}\, \exp[-A(\mathbf{X})]},
\end{aligned} \tag{7.1}
$$

and we have so far concentrated our attention on approximating such integrals using Laplace's method. The integral is evaluated in discrete time, and if we chose an interval $[t_0, t_{final}]$ over which the evaluation occurs, then indicating time in steps of Δt, we write $t_n = t_0 + n\Delta t;\ n = 0, 1, \ldots, N;\ t_N = t_{final}$. If we have paths $\mathbf{X}$ along which we integrate with D-dimensional state variables $\mathbf{x}(t_n) = \mathbf{x}(n)$ and N_p parameters $\boldsymbol{\theta}$, the dimension of the integral is $\mathcal{D} = D(N + 1) + N_p$.

We have explored and utilized Laplace's method for approximating such integrals as Eq. (7.1). This evaluates the expected value integral at the peaks of the probability distribution $P(\mathbf{X}|\mathbf{Y})$.

We now turn to another widely used and well developed method involving searching the conditional probability distribution $P(\mathbf{X}|\mathbf{Y})$ beyond the peaks, which correspond to minima in the action $A(\mathbf{X}) = -\log[P(\mathbf{X}|\mathbf{Y})]$. These are called *Monte Carlo* methods.

As the data $\mathbf{Y}$ are fixed during the integration indicated in Eq. (7.1), we search on the paths only. We will leave aside writing the $\mathbf{Y}$ for now, so $P(\mathbf{X}|\mathbf{Y}) \rightarrow P(\mathbf{X})$ unless we need to clarify the appearance of a conditional probability.

Two Monte Carlo sampling methods will be discussed (Metropolis et al. (1953); Hastings (1970); Fang et al. (2020)), but we note the generality of the methods (see sections 7.7 and 15.8 of Press et al. (2007)) and the diversity of the methods (Neal (1993, 2011); Girolami and Calderhead (2011); Ghahraman (2015); Duane et al. (1987); Haugh (2017); Betancourt (2017)).

We cannot possibly provide the reader with adequate references to the plethora of papers, books, conference proceedings, and *arXiv* entries that have been devoted to exploring and using these search methods since the early 1950s and beyond (Metropolis et al. (1953); Gubernatis (2005); Hastings (1970); Duane et al. (1987); Haugh (2017); Betancourt (2017)).

The first critical idea in Monte Carlo calculations is to choose a 'good' starting location in the space of interest, here path space. We have discussed how to choose a 'good' initial path $\mathbf{X}_{initial}$ (see Section 5.2.2) and then using that as the start of a sampling protocol. In our Precision Annealing protocol this takes place at $R_f = 0$ to get started. We call the path generated by a Monte Carlo search at this value of R_f, $\mathbf{X}_{start}$.

Then, and this is a key step, we make a **proposal** for the next place in path space after $\mathbf{X}_{start}$ to be sampled to acquire the next look at $P(\mathbf{X})$. We call this proposed path space location $\mathbf{Z}^{proposed}$.

The proposal strategy is central to the Monte Carlo procedure, and it is this operation that is the main difference between one way of sampling $P(\mathbf{X})$ and other ways of doing the sampling.

Then we require an **acceptance** criterion based on some principle and satisfying a 'detailed balance' constraint. If the acceptance criterion is met, then we move $\mathbf{Z}^{proposed}$ to $\mathbf{Z}_1^{accepted}$. We store $\mathbf{Z}_1^{accepted}$ in some convenient place and begin the "proposal, acceptance" protocol once again, starting with $\mathbf{Z}_1^{accepted}$: Namely, propose $\mathbf{Z}_1^{accepted} \rightarrow \mathbf{Z}^{proposed}$. If this is accepted, then move the new $\mathbf{Z}^{proposed} \rightarrow \mathbf{Z}_2^{accepted}$; we continue this way until we have N_A accepted paths: $\mathbf{Z}_j^{accepted}$; $j = 1, 2, \ldots, N_A$.

If the proposal $\mathbf{Z}^{proposed}$ is *rejected*, we keep the original $\mathbf{X}_{start}$ and begin the sampling search over again by making a new proposal: $\mathbf{X}_{start} \rightarrow \mathbf{Z}^{proposed}$.

The sequence $\mathbf{X}_{start} \rightarrow \mathbf{Z}^{proposed} \rightarrow \mathbf{Z}_j^{accepted}$ or back to $\mathbf{X}_{start}$, then storing the $\mathbf{Z}_j^{accepted}$; $j = 1, 2, \ldots N_A$ paths, is repeated until the number of accepted paths, N_A, is adequate:

After a number of 'burn-in' steps (Press et al. (2007)), we initiate the saving of all accepted paths $\mathbf{Z}_j^{accepted}$; $j = 1, 2, \ldots, N_A$ until we are confident we have enough accepted paths so that

$$\langle G(\mathbf{X}) \rangle = \frac{1}{N_A} \sum_{j=1}^{N_A} G(\mathbf{Z}_j^{accepted}) + \frac{\text{Numerator}(N_A)}{\sqrt{N_A}},$$

with

Numerator(N_A) = RMS Variation of $G(\mathbf{X})$ over N_A samples   (7.2)

is an accurate approximation to $\langle G(\mathbf{X}) \rangle$.

We'll consider two different Monte Carlo strategies for making proposals for moves from where we are, $\mathbf{X}_{start}$, to where we would like to go, $\mathbf{Z}^{proposed}$, to accomplish the next sampling of $P(\mathbf{X})$. We will use (1) **random changes** (Metropolis et al. (1953); Hastings (1970)) along the path $\mathbf{X}_{start}$, and (2) **structured changes** along a path in $\mathcal{D}$-dimensional path space (Fang et al. (2020)).

The idea of *detailed balance* in the sampling of the distribution $P(\mathbf{X})$ is inherited from the Statistical Physics of equilibrium for many body interactions in a solid or a gas (Metropolis et al. (1953); Gubernatis (2005)). Arriving at the distribution $P(\mathbf{X})$ should not depend on where one starts or what sequence of paths is utilized to provide adequate samples of $P(\mathbf{X})$, so this equilibrium should be invariant to applying one more step along the sequence of paths that takes us from $\mathbf{X}_{start}$ through the $\mathbf{Z}_j$'s. The transition operator moving us one step $\mathbf{Z}_{i-1} \rightarrow \mathbf{Z}_i$ along the sequence of paths we'll denote as $P_{transition}(\mathbf{Z}_i|\mathbf{Z}_{i-1})$. Invariance of $P(\mathbf{X})$ means (Haugh (2017))

$$P(\mathbf{W}) = \sum_{\mathbf{X}} P_{transition}(\mathbf{W}|\mathbf{X}) P(\mathbf{X}). \tag{7.3}$$

This is also called a stationary distribution. Such an invariant distribution is the target of the sequence of accepted paths after a transient from any $\mathbf{X}_{start}$, namely after the 'burn-in.'

The sequence or 'chain' is reversible, so we may think of it in the same way as we view the physical equilibrium distribution in a gas, which is to say,

$$P_{transition}(\mathbf{X}|\mathbf{Z}) P(\mathbf{Z}) = P_{transition}(\mathbf{Z}|\mathbf{X}) P(\mathbf{X}), \tag{7.4}$$

and this implies

$$\sum_{\mathbf{X}} P_{transition}(\mathbf{Z}|\mathbf{X}) P(\mathbf{X}) = \sum_{\mathbf{X}} P_{transition}(\mathbf{X}|\mathbf{Z}) P(\mathbf{Z}) = P(\mathbf{Z}), \tag{7.5}$$

namely our sampling strategy leads to the invariant distribution $P(\mathbf{Z})$. Equation (7.4) is called the *detailed balance* condition.

Haugh (2017) argues that detailed balance is a sufficient condition for our operations to reach the distribution we seek $P(\mathbf{X}) = \exp[-A(\mathbf{X})]$, but not a necessary condition. There are discussions of Monte Carlo methods not satisfying this detailed balance criterion (Hidemaro Suwa and Todo (2010); Sohl-Dickstein et al. (2016)).

The Markov Chain Monte Carlo (known as MCMC) method of following a sequence of accepted proposals, these constitute the Markov Chain, can lead to ergodic behavior, and may lead to an even stronger outcome than ergodicity: the chain can be *mixing* (Arnol'd (1989)). Haugh (2017) notes that the time for the sampling procedure to get a distance ϵ from the $P(\mathbf{X})$ distribution is approximately $-\log[\epsilon]$. This is good news as it says, informally, one can increase the accuracy with which one arrives near or 'on' $P(\mathbf{X})$ with only a logarithmic increase in number of steps to get there.

Before discussing the two Monte Carlo strategies, let me repeat that there are a *vast* number of books, journal articles, and papers in conference proceedings addressing the many, many aspects of Monte Carlo estimation of probability distributions, and there is no way I can or should quote all of them, so instead I take the moment to thank all the authors of those scientific and technical papers. Thank you!

7.1 Metropolis-Hastings – Random Proposals

The original Metropolis-Hastings Monte Carlo methods (Metropolis et al. (1953); Hastings (1970); Gubernatis (2005)) make use of random changes in $\mathbf{X}_{start}$ to generate proposals $\mathbf{Z}^{proposed}$. One can accomplish this by following the analysis in Hastings (1970).

First, select yet another conditional probability distribution, which is up to you. This is often called $q(\mathbf{Z}|\mathbf{X})$.

Starting from $\mathbf{X}_{start}$ draw a $\mathbf{Z}^{proposed}$ from this conditional distribution $q(\mathbf{Z}|\mathbf{X}_{start})$ – this is one of the random parts. Then define the acceptance probability

$$P_{acceptance}(\mathbf{X}_{start}, \mathbf{Z}^{proposed}) = \min\left\{1, \frac{P(\mathbf{Z}^{proposed})}{P(\mathbf{X}_{start})} \frac{q(\mathbf{X}_{start}|\mathbf{Z}^{proposed})}{q(\mathbf{Z}^{proposed})|\mathbf{X}_{start})}\right\}. \tag{7.6}$$

Set $\mathbf{Z}^{proposed} \to \mathbf{X}_{start}$ with this probability. If this proposed change is accepted, $\mathbf{Z}^{proposed}$ becomes $\mathbf{Z}_1^{accepted}$, and we repeat the proposal/acceptance step starting with a new $\mathbf{X}_{start} = \mathbf{Z}_1^{accepted}$.

If the proposal is **not** accepted, we return to the previous $\mathbf{X}_{start}$ and propose a different random path as the $\mathbf{Z}^{proposed}$.

Repeat these steps until you are ready to stop. Collect all of the N_A accepted paths $\mathbf{Z}_j^{accepted}$; $j = 1, 2, \ldots N_A$ accepted paths and store them in a convenient location. These are your samples of the target conditional probability distribution $P(\mathbf{X}|\mathbf{Y})$.

Then, as noted above,

$$\langle G(\mathbf{X}) \rangle \approx \frac{1}{N_A} \sum_{j=1}^{N_A} G(\mathbf{Z}_j^{accepted}). \tag{7.7}$$

We have just described how a general Monte Carlo strategy will proceed for a single initial path $\mathbf{X}_{start}$. In practice we would select N_I initial paths $\mathbf{X}_{start}^q$;

$q = 1, 2, \ldots, N_I$ using the initialization method described earlier but with N_I different draws for the initial choice of parameters $\boldsymbol{\theta}$ and unmeasured state variables.

We will do precision annealing, which means that at each $\beta = \log_\alpha[R_f / R_{f0}]$ (or R_f) value, we will gather into a collection $\mathbf{Z}_j^{q-accepted}$ paths for $q = 1, 2, \ldots, N_I$; $j = 1, 2, \ldots, N_A^\beta$, and we may accept a different number of paths emerging from each of the $\mathbf{X}_{start}^q$ paths. At each fixed β we will collect N_A^β accepted paths.

From these we create N_I average accepted paths at each β called $\overline{\mathbf{Z}}_\beta^q$:

$$\overline{\mathbf{Z}}_\beta^q = \frac{1}{N_A^\beta} \sum_{j=1}^{N_A^\beta} \mathbf{Z}_{\beta, j}^{q-accepted}. \tag{7.8}$$

The expected value of the function $G(\mathbf{X})$ on the N_I paths emerging from $\mathbf{X}_{start}^q$ selecting N_A^β accepted paths at each β is

$$\langle G(\mathbf{X}) \rangle \Big|_\beta = \frac{1}{N_A^\beta} \sum_{q=1}^{N_A^\beta} G(\overline{\mathbf{Z}}_\beta^q), \tag{7.9}$$

and we are interested in the following average as the approximation to the expected value $\langle G(\mathbf{X}) \rangle$:

$$\langle G(\mathbf{X}) \rangle_{\beta max} = \frac{1}{N_A^{\beta max}} \sum_{q=1}^{N_A^{\beta max}} G(\overline{\mathbf{Z}}_{\beta max}^q). \tag{7.10}$$

Several comments are in order here:

- There are many choices to be made along the trail from choosing $\mathbf{X}_{start}$, selecting $q(\mathbf{Z}|\mathbf{X})$, and eventually reaching the accepted paths $\mathbf{Z}_j^{accepted}$. Ergodicity of the collection of these accepted paths may guarantee independence of one's starting path $\mathbf{X}_{start}$. If the collection of paths is mixing (Arnol'd (1989)), this is a stronger assurance of the independence of $\langle G(\mathbf{X}) \rangle$ on $\mathbf{X}_{start}$. However, there is no clue given how long it will take in terms of total number of steps to perform that desired mixing collection.
- One can see quite clearly that if $q(\mathbf{Z}|\mathbf{X})$ is symmetric in its arguments, it cancels out of the acceptance probability,

$$P_{acceptance}(\mathbf{X}_{start}, \mathbf{Z}^{proposed}) = min\left\{ 1, \frac{\exp\left[-A(\mathbf{Z}^{proposed})\right])}{\exp[-A(\mathbf{X}_{start})]} \right.. \tag{7.11}$$

- It is possible to go 'downhill' as well as 'uphill' in the value of the action.

- In implementing the MHR-MC method, we do not have to evaluate the gradient of $A(\mathbf{X})$ in path space (Metropolis et al. (1953); Gubernatis (2005); Hastings (1970)). This is an advantage over the Laplace variational methods.
- As these methods have been adopted in many scientific fields, it should be no surprise the literature is so vast. Bonne chance in sorting through the articles, books, and proceedings of meetings concerned with Monte Carlo methods.

Because of Marshall Rosenbluth's contributions to the MHR algorithm as recounted in Gubernatis (2005), I join Gubernatis in calling it the MHR algorithm. Join in too.

We will now discuss the use of MHR Monte Carlo methods, which we will call *Random Proposals*, executed along with the Precision Annealing technique developed in Chapter 5. We will follow that with the use of Hamiltonian Monte Carlo methods, differing from MHR methods in the way proposals are made.

7.2 Precision Annealing MHR Sampling

We discussed precision annealing in an earlier chapter (Chapter 5) and outlined how one does it for variational calculations seeking the minimum of the action. The procedure in the case of Monte Carlo calculations is similar but must account for the operation of the MHR process or, later, the use of Hamilton Monte Carlo methods (Fang et al. (2020)) for making proposals.

We adopted the following annealing schedule in Precision Annealing for R_f:

$$R_f = R_{f_0}\,\alpha^{\beta}, \tag{7.12}$$

with $\alpha > 1$ and $\beta = 0, 1, \ldots, \beta_{max}$. R_{f_0} should be small. A choice of α near unity leads to the slow increase in R_f as β increases, introducing the nonlinearity of the model in an adiabatic manner. At each R_f value, $q = 1, 2, \ldots N_I$, MC calculations are performed starting from the solution generated by the procedure at the previous R_f.

7.2.1 $\beta = 0$

At $\beta = 0$, we select, without repeats, each of the N_I initial paths $\mathbf{X}_{\text{init}}^q$ identified at $R_f = 0$ to be an initial condition. (See Section 5.1.1.)

We then start to sample using the MHR procedure with $R_f = R_{f_0}$ as the starting hyperparameter. We collect $N_A^{\beta=0}$ accepted paths for each initial condition ensemble member. These we denote by $\mathbf{Z}_{\beta=0,\,j}^{q-accepted}$ with $j = 1, \ldots, N_A^{\beta=0}$, which are generated along the Markov chain.

Next, we form the sample mean of each of the N_I Markov chains developed so far,

$$\overline{\mathbf{Z}}^q_{\beta=0} = \frac{1}{N^{\beta=0}_A} \sum_{j=1}^{N^{\beta=0}_A} \mathbf{Z}^{q-accepted}_{\beta=0,\,j} \,, \tag{7.13}$$

as the initial path for the next β value.

Since the N_I calculations are done independently and in parallel, this results in a set of N_I averages over accepted paths $\overline{\mathbf{Z}}^q_{\beta=0}$ with $q = 1, \dots, N_I$.

7.2.2 $\beta = 1$

At the next step we move to $\beta = 1$. We select N_I initial conditions: $\overline{\mathbf{Z}}^q_{\beta=0}$ with $q = 1, \dots, N_I$, available for our HMC evaluations. For each initial condition we collect $N^{\beta=1}_A$ sampled paths and form the N_I means for each:

$$\overline{\mathbf{Z}}^q_{\beta=1} = \frac{1}{N^{\beta=1}_A} \sum_{j=1}^{N^{\beta=1}_A} \mathbf{Z}^{q-accepted}_{\beta=1,\,j} \,. \tag{7.14}$$

These are N_I initial conditions for the next value of β.

7.2.3 $\beta = 2, \dots, \beta_{max}$

We continue this procedure until we reach a useful β_{max}. By plotting two quantities, the action and the model error, versus R_f or $\beta = \log_\alpha \frac{R_f}{R_m}$ we will have insight on how to select this.

First, we should make a plot of the action determined by each of the N_I paths at each R_f as a function of β. We find that the action becomes independent of β as the precision of the model increases. The origin of this is seen in a second graphic where we plot the model error term in the action:

$$\sum_{n=0}^{N-1} \sum_{a=1}^{D} \frac{R^{x_a}_f}{2N} \left[x_a(n+1) - f_a(\mathbf{x}(n), \boldsymbol{\theta}) \right]^2 , \tag{7.15}$$

as a function of β.

If the model is consistent with the data, the model error will rapidly decrease to a small value as β increases. The action is the sum of the model error, the measurement error, and the initial condition, as the model error tends numerically to a small number, the action levels out to the sum of the measurement error and the initial condition, which are both independent of R_f. Indeed, the measurement error term provides a lower bound to the action and indicates whether the selected model is "right." This independence of the action with respect to β will also depend in SDA on the number of measurements L at each time observations are made.

As a result of these MHRMC calculations, the expected value of functions $G(\mathbf{X})$ in the original space, Eq. (7.1), will be estimated as

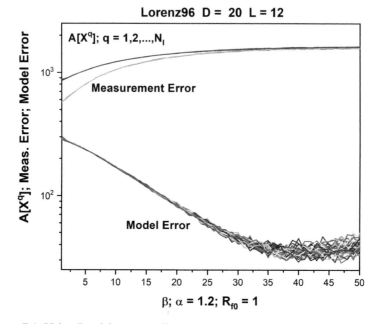

Figure 7.1 Using Precision Annealing (PA) MHR Monte Carlo methods we evaluated the action (Eq. (7.17)), the measurement error term, and the Model Error
term for the Lorenz96 dynamics (Eq. (7.18)). In the PA protocol we used $\alpha = 1.2$
and $R_{f0} = 1.0$. We evaluated the action, the measurement error, and the model
error for $\beta = \log_{\alpha}[R_f/R_{f0}]$ from $\beta = 0$ to $\beta = 50$, and we selected $N_I = 50$
initial paths. The initialization of each path was as described in Section 5.6.1. New
paths were proposed by randomly changing a small subset of the parameters and
all the state variables at each step in β. The dimension D of the Lorenz96 model
was 20, and the number of observations at each measurement time was $L = 12$.
Because the model error becomes a few percent of the measurement error, the
action essentially levels off as R_f becomes large.

$$\langle G(\mathbf{X}) \rangle \approx \frac{1}{N_I} \sum_{q=1}^{N_I} G\left(\overline{\mathbf{Z}}_{\beta_{\max}}^{(q)}\right). \tag{7.16}$$

7.2.4 Lorenz96; D = 20; PAMHR Method

Using these methods for MHR sampling with Precision Annealing, we take the
action

$$A(\mathbf{X}) = \sum_{k=0}^{F} \sum_{\ell=1}^{L} \frac{R_m}{2(F+1)} \left[x_\ell(\tau_k) - y_\ell(\tau_k) \right]^2$$
$$+ \sum_{n=0}^{N-1} \sum_{a=1}^{D} \frac{R_f^{x_a}}{2N} \left[x_a(n+1) - f_a(\mathbf{x}(n), v) \right]^2, \tag{7.17}$$

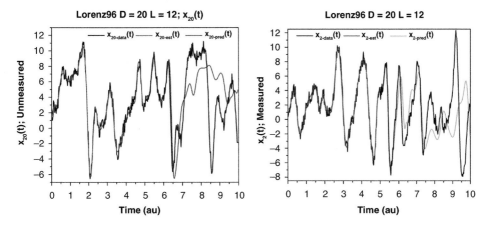

Figure 7.2 In the PAMHR protocol we used $\alpha = 1.2$ and $R_{f0} = 1.0$. We evaluated the action, the measurement error, and the model error for $\beta = \log_\alpha[R_f/R_{f0}]$ from $\beta = 0$ to $\beta = 50$, and we selected $N_I = 50$ initial paths. The initializations of the path were as described in Section 5.6.1. New paths were proposed by randomly changing a small subset of the parameters and all the state variables at each step in β. The dimension D of the Lorenz96 model was 20, and the number of observations at each measurement time was $L = 12$. **Left Panel** Time series of the unmeasured state variable $x_{20}(t)$ in the Lorenz96 dynamical equations. We show the noisy data in black, the estimated $x_{20}(t)$ in red during the observation window $0 \leq t \leq 5$, and the predicted value of $x_{20}(t)$ in blue within the time window $5 \leq t \leq 10$. The departure of the predictions from the known values of the data comes from the chaotic behavior of the dynamical equations at $\nu = 8.17$. **Right Panel** Time series of the unmeasured state variable $x_2(t)$ in the Lorenz96 dynamical equations. We show the noisy data in black, the estimated $x_2(t)$ in red during the observation window $0 \leq t \leq 5$ and the predicted value of $x_2(t)$ in green within the time window $5 \leq t \leq 10$. The departure of the predictions from the known values of the data comes from the chaotic behavior of the dynamical equations at $\nu = 8.17$. The very good predictions for $t \geq 5$ indicate that the full state of the dynamical system (Eq. (7.18)) has been accurately estimated at $t = 5$. Predictions were made using $\mathbf{x}(t = 5)$ as the initial condition and the estimated value of ν.

with $\mathbf{f}(\mathbf{x}(n), \nu)$ as a time discretization for the Lorenz96 model whose dynamical equations we recall:

$$\frac{dx_a(t)}{dt} = x_{a-1}(t)(x_{a+1}(t) - x_{a-2}(t)) - x_a(t) + \nu. \tag{7.18}$$

We require $x_{-1}(t) = x_{D-1}(t)$; $x_0(t) = x_D(t)$; $x_1(t) = x_{D+1}(t)$. ν is a fixed time independent parameter. We take $\nu = 8.17$. The solutions to Eq. (7.18) are chaotic for this choice of ν (Lorenz and Emanuel (1998); Kostuk (2012)). We are using ν as the parameter in the Lorenz96 equations here.

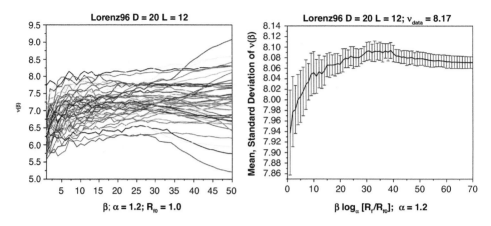

Figure 7.3 Using PAMHR to estimate the parameter ν in the Lorenz96 dynamics (Eq. (7.18)), D = 20, we estimated $N_I = 30$ values of $\nu(\beta)$ at each β. These are shown in the **Left Panel**. In the **Right Panel** we show the average estimated value of $\nu(\beta)$ along with the estimated RMS error at each β.

We performed a twin experiment using as 'data' a fourth order Runge-Kutta solution to the Lorenz96 equations, D = 20, with $\Delta t = 0.025$ and Gaussian noise $\mathcal{N}(0, 0.25)$ added to each state variable $\mathbf{x}(t)$. This means $R_m = 4.0$.

7.3 Hamiltonian Monte Carlo Methods – Structured Proposals

7.3.1 General Idea of Hamiltonian Monte Carlo

Hamiltonian Monte Carlo (HMC) (Duane et al. (1987); Neal (2011); Betancourt (2017); Haugh (2017)) strikes out in a new direction. It adds apparent 'auxiliary' degrees-of-freedom, called π, to the $\mathcal{D}$-dimensional path $\mathbf{X}$. The sampling of the conditional probability distribution is done in $\{\mathbf{X}, \pi\}$ 'phase' space in a way that preserves some invariants of the rules for moving about in that space **and** maintains the expected values (Eq. (7.1)), as we will see.

Proposals are made in phase space moving from $\{\mathbf{X}_{start}, \pi_{start}\}$ to another location in phase space $\{\mathbf{X}^{proposed}, \pi^{proposed}\}$ using rules that preserve a scalar quantity, which stands in for the action **and** preserves the phase space volume, $d\mathbf{X}\,d\pi$. This strategic insight permits large jumps in the original path space $\mathbf{X}$ while keeping the acceptance probability near unity. The efficiency in sampling the target conditional probability distribution is much larger than MHR with Random Proposals (Fang et al. (2020)). As introduced in Duane et al. (1987) the added variables π are selected to be *canonically conjugate momenta*, familiar to physicists in Hamiltonian mechanics (Goldstein et al. (2002); Arnol'd (1989)) and thoroughly analyzed for about two centuries. By choosing the additional variables π to be canonical conjugates of $\mathbf{X}$, one can use the rules of classical mechanics to move around in $(\mathbf{X}, \pi)$ phase space and preserve the underlying symplectic structure in making every proposal (Gelfand and Fomin (1963); Arnol'd (1989); Hairer et al. (2006); Goldstein et al. (2002)).

We have discussed how this is accomplished both when the motion in $(\mathbf{X}, \pi)$ phase space is labeled by a continuous variable we call s, and how we should modify this when we take discrete steps Δs in that label.

The Hamiltonian dynamics determining the motion, from a chosen starting location $\{\mathbf{X}, \pi\}$ to another phase space location, $\{\mathbf{X}', \pi'\}$: $\{\mathbf{X}, \pi\} \rightarrow \{\mathbf{X}', \pi'\}$ is generated by a scalar $H(\mathbf{X}, \pi)$ called, with no surprises here, the Hamiltonian. In the motion, labeled by a scalar called s and referred to as 'time,' through $\{\mathbf{X}, \pi\}$ phase space, we utilize

$$\frac{dA(\mathbf{X}(s), \pi(s))}{ds} = \left\{ A(\mathbf{X}(s), \pi(s)), H(\mathbf{X}(s), \pi(s)) \right\} \bigg|_{\mathbf{X}(s), \pi(s)}, \qquad (7.19)$$

where $A(\mathbf{X}, \pi)$ is any function of the variables $(\mathbf{X}, \pi)$, and Eq. (7.19) is the Poisson bracket. The Hamiltonian is conserved along trajectories labeled by s as is volume in phase space: $d\mathbf{X}\,d\pi = d\mathbf{X}'\,d\pi'$. Many other conserved quantities exist (Goldstein et al. (2002); Arnol'd (1989)), but they generally have no relation to the discussion here.

The idea of Duane et al. (1987) was to add to the action $A(\mathbf{X})$ an additional function of π alone, $h(\pi)$, to form a Hamiltonian $H(\mathbf{X}, \pi) = h(\pi) + A(\mathbf{X})$.

In the formulation of the expected values of functions $G(\mathbf{X})$ of $\mathbf{X}$ alone,

$$\langle G(\mathbf{X}) \rangle = \frac{\int d\mathbf{X} G(\mathbf{X}) \exp[-A(\mathbf{X})]}{\int d\mathbf{X} \exp[-A(\mathbf{X})]}, \qquad (7.20)$$

introducing an $h(\pi)$ into this integral in a judicious manner results in:

$$\langle G(\mathbf{X}) \rangle = \frac{\int d\mathbf{X} \, G(\mathbf{X}) \exp[-A(\mathbf{X})]}{\int d\mathbf{X} \exp[-A(\mathbf{X})]},$$

$$\langle G(\mathbf{X}) \rangle = \frac{\int d\mathbf{X} \, G(\mathbf{X}) \exp[-A(\mathbf{X})]}{\int d\mathbf{X} \exp[-A(\mathbf{X})]} \frac{\int d\pi \, \exp[-h(\pi)]}{\int d\pi \, \exp[-h(\pi)]},$$

$$\langle G(\mathbf{X}) \rangle = \frac{\int d\mathbf{X} d\pi \, G(\mathbf{X}) \exp[-A(\mathbf{X}) - h(\pi)]}{\int d\mathbf{X} \, d\pi \, \exp[-A(\mathbf{X}) - h(\pi)]},$$

$$\langle G(\mathbf{X}) \rangle = \frac{\int d\mathbf{X} d\pi \, G(\mathbf{X}) \exp[-H(\mathbf{X}, \pi)]}{\int d\mathbf{X} \, d\pi \, \exp[-H(\mathbf{X}, \pi)]}. \qquad (7.21)$$

This tells us that the expected values $\langle G(\mathbf{X}) \rangle$, which are the goals of SDA and machine learning (Chapter 8), are unchanged under HMC protocols for an arbitrary choice of the canonical momenta and the function $h(\pi)$.

Making proposals in phase space enables large jumps in $\mathbf{X}$ path space while remaining on the surface $H(\mathbf{X}, \pi) = $ constant. Such proposals will also result in

$$P_{acceptance}(\mathbf{X}_{start}\pi_{start}, \mathbf{Z}^{proposed}, \pi^{proposed}) = 1, \quad (7.22)$$

since $H(\mathbf{X}, \pi)$ is constant. This consequence is correct when proposals $\{\mathbf{X}(0), \pi(0)\} \rightarrow \{\mathbf{X}(s), \pi(s)\}$ are made in *continuous* 'time' s moving from $s = 0$ to s using the rules of classical Hamiltonian mechanics:

$$\frac{d\mathbf{X}(s)}{ds} = \frac{\partial H(\mathbf{X}(s), \pi(s))}{\partial \pi(s)}; \quad \frac{d\pi(s)}{ds} = -\frac{\partial H(\mathbf{X}(s), \pi(s))}{\partial \mathbf{X}(s)}. \quad (7.23)$$

Of course, we cannot move along in continuous time in a numerical strategy.

One must distinguish the use of the scalar label s in HMC from the time labels t and τ as described in Chapter 2. s is just a label to track movements in $\{\mathbf{X}, \pi\}$ phase space. Actual times t and τ are labels used in identifying the path of model dynamical variables $\mathbf{x}(t)$ and the data $\mathbf{y}(\tau)$ throughout an observation window $[t_0, t_{final}]$.

7.3.2 HMC in Discrete 'time' 's'

The HMC procedure (Duane et al. (1987); Neal (2011); Betancourt (2017)) doubles the dimension of path space $\mathbf{X}$, introducing a physics-motivated but arbitrarily chosen canonical momentum π associated with the path space position $\mathbf{X}$.

Many choices for the "kinetic energy," $h(\pi)$, are useful, but the selection of $h(\pi) = \pi \cdot \pi/2$ is convenient, and the resulting distribution $\exp[-h(\pi)]$ is Gaussian. This is easy to sample (Press et al. (2007)).

Kadakia (2016) has pointed out that small modifications of this standard choice to include higher order than quadratic polynomials of π can introduce chaotic behavior and substantial mixing in motions generated by $H(\mathbf{X}, \pi)$.

With this choice the Hamiltonian $H(\mathbf{X}, \pi)$ is:

$$H(\mathbf{X}, \pi) = A(\mathbf{X}) + h(\pi) = A(\mathbf{X}) + \frac{\pi \cdot \pi}{2}, \quad (7.24)$$

and as we saw in Eq. (7.21) this replaces $A(\mathbf{X})$.

The core idea of HMC (Fang et al. (2020); Betancourt (2017)) is to propose $(\mathbf{X}(s), -\pi(s))$ using Eq. (7.23) starting from $(\mathbf{X}(0), \pi(0))$.

This is precise when the integration is performed with s taken as a continuous variable, a Hamiltonian flow is thus realized and $H(\mathbf{X}, \pi)$ is conserved. A complete HMC proposal includes a negative sign before $\pi(s)$, meaning that we need to flip the momentum *after* the Hamiltonian integration is finished. This flipping is to make the proposal reversible in s and symmetric in $(\mathbf{X}, \pi)$ as described by Betancourt (2017). This assures detailed balance. Because π appears only as $\pi \cdot \pi$ in $h(\pi)$, changing $\pi \rightarrow -\pi$ does not change the acceptance probability.

In practice, the nonlinear terms in the model error element of $A(\mathbf{X})$ preclude analytical evaluations of the expected value integrals Eq. (7.1), and one must work with discrete s in order to integrate Eq. (7.23). When we use a *symplectic integrator* to perform the integration of Hamilton's equations in discrete s, a result of Ge and Marsden Zhong and Marsden (1988) tells us that we cannot precisely preserve both the symplectic form *and* $H(\mathbf{X}, \pi)$ at the same time. Therefore, we anticipate that by using a discrete s symplectic integrator, we will find $H(\mathbf{X}(s), \pi(s)) \approx H(\mathbf{X}(0), \pi(0))$, but not exactly equal. As a consequence, when determining the acceptance or rejection of the proposal $(\mathbf{X}(s), \pi(s))$, the acceptance probability can usually be near unity, but not precisely unity, that is,

$$P_{acceptance}(\{\mathbf{X}(s), -\pi(s)\}; \{\mathbf{X}(0), \pi(0)\})$$

$$= \min\left\{1, \frac{\exp\left[-H(\mathbf{X}(s), -\pi(s))\right]}{\exp\left[-H(\mathbf{X}(0), \pi(0))\right]}\right\} \approx 1. \tag{7.25}$$

As the parameters $\boldsymbol{\theta}$ in SDA and machine learning are taken to be components of the vector $\mathbf{X}$, they also have conjugate variables $\boldsymbol{\eta}$ included in π. HMC provides a principled manner of exploring $\mathbf{P}(\mathbf{X}|\mathbf{Y})$ on both state variables and time independent parameters.

7.3.3 PAHMC

In addition to using HMC to make proposals $\{\mathbf{X}(0), \pi(0)\} \rightarrow \{\mathbf{X}(s), -\pi(s)\}$ in phase space, when implementing HMC, one must select a symplectic integrator in order to preserve the symplectic invariants (Neal (2011); Hairer et al. (2006)), including $H(\mathbf{X}, \pi)$ and $d\mathbf{X}d\pi$, many choices are available. We have used a fairly common choice called the *leapfrog symplectic integrator* (Blanes et al. (2014)), which possesses simplicity and accuracy. To begin sampling, we initialize the Hamiltonian mechanics at $s = 0$ using the initialization procedure described in Chapter 5 at $R_f = 0$.

Performing a selection of $j = 1, 2, \ldots, N_I$ proposal steps achieved by integrating the Hamilton dynamical equations (Eq. (7.23)): $\{\mathbf{X}_j(0), \pi_j(0)\} \rightarrow \{\mathbf{X}_j(s), -\pi_j(s)\}$. This first Hamiltonian mechanics step yields $j = 1, 2, \ldots, N_I$ starting paths $\mathbf{X}_{start-j}$.

We then move $R_f \rightarrow R_{f0}$, equivalently we move to $\beta = 0$, and the momentum $\pi_{start-j}$ is selected by sampling from $\exp[-h(\pi)] = \exp[-\pi^2/2]$ to select $\pi_{start-j} = \pi(0)$ and choosing $\mathbf{X}_j(0)$ to be the last accepted 'configuration space' paths $\mathbf{X}_j \; j = 1, 2, \ldots, N_I$ using Eq. (7.25).

We first move $(\mathbf{X}(0), \mathbf{P}(0))$ to $(\mathbf{X}(\epsilon), \boldsymbol{\pi}(\epsilon))$ selecting the 'leapfrog' symplectic stepping rule,

$$
\begin{cases}
\boldsymbol{\pi}(\epsilon/2) = \boldsymbol{\pi}(0) - \dfrac{\epsilon}{2} \nabla A(\mathbf{X}(0)), \\[2mm]
\mathbf{X}(\epsilon) = \mathbf{X}(0) + \epsilon\, \boldsymbol{\pi}(\epsilon/2), \\[2mm]
\boldsymbol{\pi}(\epsilon) = \boldsymbol{\pi}(\epsilon/2) - \dfrac{\epsilon}{2} \nabla A(\mathbf{X}(\epsilon)),
\end{cases}
\tag{7.26}
$$

so $\epsilon = \Delta s$.

Then, starting from $(\mathbf{X}(\epsilon), \boldsymbol{\pi}(\epsilon))$, we move to the next point at $(\mathbf{X}(2\epsilon), \boldsymbol{\pi}(2\epsilon))$. After proceeding for S steps, we arrive at a proposal (Betancourt (2017); Haugh (2017)) $(\mathbf{X}_j^{proposed}, \boldsymbol{\pi}_j^{proposed}) = (\mathbf{X}(S\epsilon), -\boldsymbol{\pi}(S\epsilon))$. These proposals $(\mathbf{X}_j^{proposed}, \boldsymbol{\pi}_j^{proposed})$ are accepted or rejected according to Eq. (7.25).

7.4 Underappreciating HMC

Much of the literature addressing HMC in a Machine Learning context attends to modifications of directly integrating Hamilton's equations (Eq. (7.23)), and these suggestions are satisfied when the rule leaves $H(\mathbf{X}, \boldsymbol{\pi})$ unchanged.

The preservations of phase space volume and the other symplectic invariants are often untouched and underappreciated. Conservation of phase space volume is essential in the use of HMC in the evaluation of expected values.

7.5 Using PAHMC on Two Model Dynamics

To illustrate the outcome of using PAHMC we apply it to noisy twin experiment 'data' from two models we have used before: (a) a Hodgkin-Huxley neuron model (NaKL) and (b) the Lorenz96 model with D = 20 and various values of L.

7.6 HH NaKL Model

To illustrate the effectiveness of the PAHMC method, we return to the twin experiment on the NaKL Hodgkin-Huxley model.

This HH model has four state variables $\{V(t), m(t), h(t), n(t)\}$. The first of these is the membrane voltage, and it is observable. $V(t)$ has a dynamical range from about -80 mV to about $+40$ mV. The final three are the voltage dependent opening and closing probabilities for Na^+ ion channels $\{m(t), h(t)\}$ and the voltage dependent opening probability for the K^+ channel $\{n(t)\}$. The gating variables $a(t) = \{m(t), h(t), n(t)\}$ all have a range $0 \le a(t) \le 1$, and they are unobservable in experiments (as of 2020). In the notation we have been using $D = 4$, $L = 1$, $N_p = 18$.

The time evolution of the states are governed by the following differential equations:

$$C_m \frac{dV(t)}{dt} = I_{inj}(t) + g_{Na} m(t)^3 h(t) \left(E_{Na} - V(t) \right)$$

$$+ g_K n(t)^4 \left(E_K - V(t) \right) + g_L \left(E_L - V(t) \right)$$

and

$$\frac{da(t)}{dt} = \frac{a_0(V(t)) - a(t)}{\tau_a(V(t))}, \tag{7.27}$$

where $I_{inj}(t)$ is the external stimulus.

We select $C_m = 1.0$, and the $a(t)$ satisfy the first order kinetics with:

$$a_0(V(t)) = \frac{1}{2} + \frac{1}{2} \tanh \left(\frac{V(t) - V_a}{\Delta V_a} \right),$$

$$\tau_a(V(t)) = \tau_{a0} + \tau_{a1} \left[1 - \tanh^2 \left(\frac{V(t) - V_a}{\Delta V_a} \right) \right]. \tag{7.28}$$

$0 \le a(t) \le 1$.

The functional dependence of $a_0(V)$ and $\tau(V)$ is derived from the very helpful database on experimental observations of Na^+ and K^+ ion channels at the Yale University *Senselab* database (Senselab-Yale (2020)). Other functional forms for these quantities are used in the literature; all are phenomenological, utilized in one laboratory experiment or another, and not set by biophysical principles. We suggest the user choose one that works for them and stick to it.

$\mathbf{f}(\mathbf{x}(k), \boldsymbol{\theta})$ denotes the model vector field in discrete time. $D = 4$ and $L = 1$. $\mathbf{x}(t) = \{x_1(t) = V(t), x_2(t) = m(t), x_3(t) = h(t), x_4(t) = n(t)\}$ and $\boldsymbol{\theta}$ is the collection of model parameters. We can measure only $x_1(t) = V(t) = V_{data}(t) = y(t)$ and the standard action can be written as

$$A(\mathbf{X}) = \frac{R_m}{2M} \sum_{k=1}^{M} \left(x_1(k) - y(k) \right)^2 + \sum_{a=1}^{4} \sum_{k=1}^{M-1} \frac{R_f^{(x_a)}}{2M} \left(x_a(k+1) - f_a(\mathbf{x}(k), \boldsymbol{\theta}) \right)^2. \tag{7.29}$$

Choosing $\Delta t = 0.02$ ms; $M = 5000$; $t_{final} = 100$ ms is the length of the observation window. $R_{f0}^{(V)} = 0.1$, $R_{f0}^{(m)} = 1200$, $R_{f0}^{(h)} = 1600$, and $R_{f0}^{(n)} = 2100$.

Data was generated using the trapezoidal rule, Eq. (7.30),

$$f_a(\mathbf{x}(m), v) = x_a(m) + \frac{\Delta t}{2} [\mathbf{F}_a(\mathbf{x}(m+1), v) + \mathbf{F}_a(\mathbf{x}(m), v)], \tag{7.30}$$

where $\mathbf{F}_a(\mathbf{x}, v)$ is the model vector field. Gaussian noise of $\mathcal{N}(0, 1)$ was then added to $x_1(t_k)$ and stored as $y(t) = V_{data}(t)$. Only $V_{data}(t)$ was presented to the model, so $L = 1$.

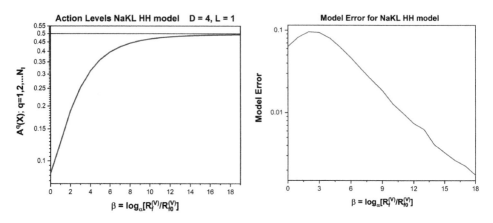

Figure 7.4 Action Levels and Model Error in PAHMC protocol for the NaKL HH Model. **Left Panel** Action Levels versus $\beta = \log_\alpha[\frac{R_f^{(V)}}{R_{f0}^{(V)}}]$. $\alpha = 2$; $\beta = 0, 1, \ldots, 19$. **Right Panel** Model Error versus β; $\alpha = 2$. The Action Levels become independent for $\beta \approx 15$ and larger indicating the model is very well satisfied as at that β the model error is several orders of magnitude smaller than the measurement error.

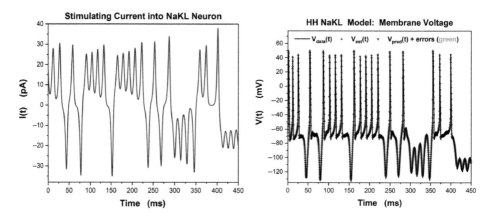

Figure 7.5 Stimulating Current and Voltage Response for HH NakL Neuron: Eq. (7.27) and Eq. (7.28) using the PAHMC protocol. **Left Panel** Stimulating Current, $I_{\text{inj}}(t)$, presented to the HH NaKL model. **Right Panel** Membrane Voltage Response to this Stimulating Current. These are the data used in estimating $V(t)$ and the voltage gated variables $\{m(t), h(t), n(t)\}$ and the $N_p = 18$ parameters reported in the Table. We set $C_m = 1.0$. 450 ms of Stimulating Current and Voltage response is shown; 100ms was used in the data assimilation information transfer. 350ms is a prediction window for $V(t)$. The noisy data is in black; the estimated voltage [0 ≤ t ≤ 100ms] is shown in red; the predicted voltage [100ms ≤ t ≤ 450ms] is shown in blue. The prediction errors are shown in green.

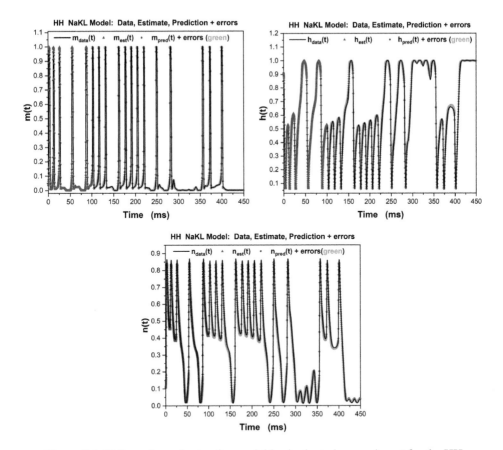

Figure 7.6 Voltage dependent gating variables in the twin experiment for the HH NaKL model. **Top Left Panel** Na activation variable m(t). **Top Right Panel** Na inactivation variable h(t). **Bottom Panel** K inactivation variable n(t). 100ms of V(t) and $I_{inject}(t)$ were used in the data assimilation information transfer. 350ms is a prediction window for $a(t) = \{m(t), h(t), n(t)\}$. The noisy data is in black; the estimated $a(t)$ [$0 \leq t \leq 100$ms] is shown in red; the predicted $a(t)$ [100ms $\leq t \leq 450$ms] is shown in blue. The prediction errors are shown in green.

7.7 Lorenz96 Model; $D = 20$

The action is our standard model as we have used before:

$$A(\mathbf{X}) = \sum_{k=0}^{F} \sum_{\ell=1}^{L} \frac{R_m}{2(F+1)} \left[x_\ell(\tau_k) - y_\ell(\tau_k) \right]^2$$

$$+ \sum_{m=0}^{M-1} \sum_{a=1}^{D} \frac{R_f}{2M} \left[x_a(m+1) - f_a(\mathbf{x}(m), v) \right]^2. \quad (7.31)$$

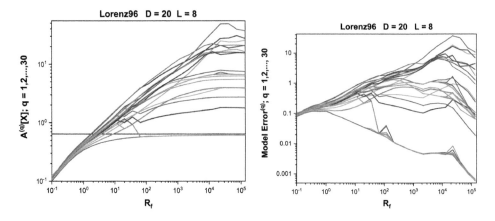

Figure 7.7 **Left Panel** The Action Levels versus R_f for the Lorenz96 model with $D = 20$ and $L = 8$ of the dynamical variables $\mathbf{x}(t)$ observed. Each step in the Precision Annealing procedure is associated with one value of R_f. We perform $N_I = 30$ PAHMC calculations, starting at N_I different $(\mathbf{X}(0), \boldsymbol{\pi}(0))$ and resulting in the paths whose action levels are shown here. $N_I = 30$ independent calculations are performed at each R_f, resulting in the many action curves displayed. All the action levels are above the anticipated value, the value of the measurement error term of the total action, shown as the solid black line. **Right Panel** The model errors in Eq. (7.31) as a function of R_f for the Lorenz96 model with $D = 20$ and $L = 8$. $N_I = 30$ independent calculations are present. At smaller R_f values, the model error dominates the action, indicating that the HMC calculations have not yet found a path that agrees with the model. At the final stage, large R_f, the model error starts to decrease exponentially and the paths proposed by HMC start to agree with the model.

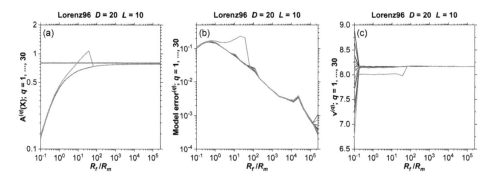

Figure 7.8 **Left Panel** The Action Levels for the Lorenz96 Model. $D = 20$, $L = 10$. **Middle Panel** The Model Error for the Lorenz96 Model. $D = 20$, $L = 10$. **Right Panel** The forcing parameter for the Lorenz96 Model. $D = 20$, $L = 10$. The true forcing parameter is $\nu = 8.17$ (solid black line).

Table 7.1 **Table of parameters θ in the HH NaKL model.** *The membrane capacitance C_m has been set to 1.0. Eighteen parameters are estimated. In the Table we show the name of the parameter, the value used in generating the data from Eq. (7.27), and the estimated value using the PAHMC protocol. In initializing the PAHMC data assimilation we drew each parameter from a uniform distribution over the range shown in the right column of the Table.*

Parameter Name; θ	θ in Data	Estimated θ	Initial Range
g_{Na}	120.0	116.19	$[100,150]$
E_{Na}	50.0	49.72	$[48,52]$
g_K	20.0	21.63	$[15,25]$
E_K	-77.0	-77.05	$[(-80),(-70)]$
g_L	0.3	0.30	$[0.1,0.5]$
E_L	-54.4	-54.27	$[(-55),(-50)]$
V_m	-40.0	-40.18	$[(-45),(-35)]$
ΔV_m	15.0	14.80	$[10,20]$
τ_{m0}	0.1	0.10	$[0.01,0.2]$
τ_{m1}	0.4	0.40	$[0.3,0.5]$
V_h	-60.0	-59.34	$[(-65),(-55)]$
ΔV_h	-15.0	-14.09	$[(-20),(-10)]$
τ_{h0}	1.0	0.99	$[0.5,1.5]$
τ_{h1}	7.0	8.78	$[5,15]$
V_n	-55.0	-53.92	$[(-60),(-40)]$
ΔV_n	30.0	31.76	$[25,35]$
τ_{n0}	1.0	1.03	$[0.5, 1.5]$
τ_{n1}	5.0	4.94	$[2,10]$

The first term on the right in Eq. (7.31) is the measurement error, and the second, the model error. The trapezoidal rule is used to discretize the Lorenz96 equations of motion (Eq. (7.18)).

7.8 Computational Considerations for PAHMR and PAHMC Procedures

Both MHR and HMC are widely used in a broad variety of calculations ranging from numerical weather prediction to machine learning tasks (Goodfellow et al. (2016)). We give some estimates on the scaling of the computation as the dimension of the model and the number of time steps or layers (see Chapter 8) increases.

The complexity of one step of HMC using leapfrog integration requires evaluating $\nabla A(\mathbf{X})$ once. It, therefore, has dominant time complexity of $O(DM)$, where, as usual, D is the dimension of the underlying dynamical system and M is the number of discrete time steps in the observation window. It has been observed (Creutz (1988); Neal (2011); Mangoubi and Vishnoi (2018)) that the computation typically

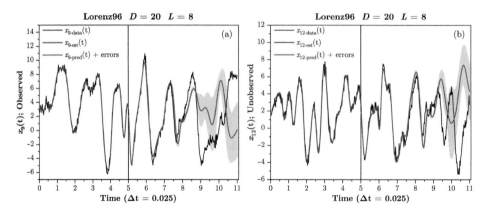

Figure 7.9 **Left Panel** Display of results for an *observed* dynamical variable $x_9(t)$ of the Lorenz96 model with $D = 20$, $L = 8$ using the PAHMC method to make proposals. The noisy data lies in the time interval $0 \le t \le 11$. The estimated values in the observation window $0 \le t < 5$ are shown in red. The predicted values are shown in blue and the prediction errors are shown in magenta in $5 < t \le 11$. **Right Panel** Display of results for an *unobserved* dynamical variable $x_{12}(t)$ of the Lorenz96 model with $D = 20$, $L = 8$. The true data, with noise added, span the full $0 \le t \le 11$ interval. The estimated values in the observation window $0 \le t < 5$ are shown in red. The predicted values, with errors (in magenta), are shown in blue within the interval of $5 < t \le 11$. For the unobserved variables, the data within $0 \le t \le 11$ are not presented to the PAHMC method

grows as the 5/4 power of the dimensions of the model, given a constant acceptance rate.

The computational burden of PAHMC grows as $O((DM)^{5/4})$. There is a multiplicative constant that is determined by the choices of the number of samples N_β and the number of precision annealing steps β_{max}.

The mixing time itself is hard to estimate, yet it has been empirically tested in many cases that HMC achieves faster mixing than other well-known Monte Carlo methods. An elegant modification of the term $h(\pi, \mathbf{X})$ in the Hamiltonian has been made by Kadakia (2016) that may further improve the mixing rate.

To compare the Random Proposals method (MHR) and HMC on the same problem we report the time for the Lorenz96 system. In Lorenz96, we are dealing with a 4000-dimensional system in path space for which $D = 20$ and $M = 200$. To obtain one HMC proposal $\mathbf{Z}^{proposed}$, we simulate the discrete Hamiltonian dynamics given in Eq. (7.26) for $S = 50$ steps with step size $\epsilon = 10^{-3}$. In an implementation in MATLAB, the computation time is around 0.02s. We then make $N_\beta = 10^3$ proposals for every β up to $\beta_{max} = 30$. The entire computation takes about 10 minutes to complete.

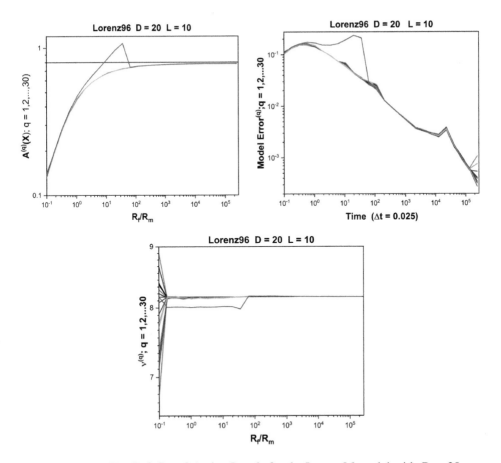

Figure 7.10 **Top Left Panel** Action Levels for the Lorenz96 model with $D = 20$ and $L = 10$ observed dynamical variables $\mathbf{x}(t)$. Each step in the Precision Annealing procedure is associated with one value of R_f. We perform $N_I = 30$ Hamiltonian Monte Carlo calculations, starting at N_I different $(\mathbf{X}(0), \boldsymbol{\pi}(0))$ and resulting in the paths shown here. $N_I = 30$ independent calculations are performed at each R_f, resulting in the action curves shown. **Top Right Panel** The model errors in Eq. (7.31) as a function of R_f for the Lorenz96 model with $D = 20$ and $L = 10$. $N_I = 30$ independent calculations are present. At smaller R_f values, the model error dominates the action. After that, the model error rapidly decreases as R_f grows. As a result, the measurement error term in Eq. (7.31) dominates the action and it becomes essentially independent of R_f. **Bottom Panel** The estimations of the Lorenz96 model forcing parameter ν at each value of R_f. $N_I = 30$ independent calculations are performed. The true forcing parameter in the twin experiments is $\nu = 8.17$ (solid black line). With $D = 20$ and $L = 10$, the correct forcing parameter quickly emerges from the random initialization.

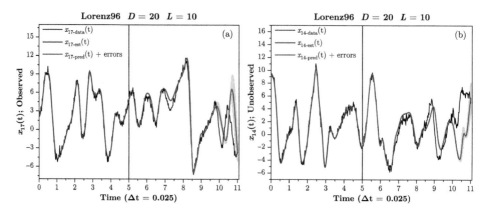

Figure 7.11 Using PAHMC for the Lorenz96 model. **Left Panel** Display of results for an observed dynamical variable $x_{17}(t)$ of the Lorenz96 model with $D = 20$, $L = 10$. The data, with noise added, span the full $0 \leq t \leq 8$ interval; they are displayed in black. The estimated values in the observation window $0 \leq t < 5$ are shown in red. The predicted values, with errors (in magenta), are shown in blue within the interval of $5 < t \leq 8$. **Right Panel** Display of results for an unobserved dynamical variable $x_{14}(t)$ of the Lorenz96 model with $D = 20$, $L = 10$. The true data, with noise added, span the full $0 \leq t \leq 11$ interval. The estimated values in the observation window $0 \leq t < 5$ are shown in red. The predicted values, with errors (in magenta), are shown in blue within the interval of $5 < t \leq 8$. For the unobserved variables, the data within $0 \leq t \leq 8$ are not presented to the PAHMC method.

A direct comparison to the RP HMC approach, we have run the same calculations for the Lorenz 96 system, $D = 20$ and $M = 200$. At each β up to $\beta_{max} = 50$, we make 6000 perturbations on $\mathbf{X}$. This means we make 2.4×10^7 proposals for each β, given that we only perturb one component of $\mathbf{X}$. As realized in the programming language C, this calculation took about four hours to complete. The computation time is about 10 times longer than the equivalent PAHMC method. The MHR approach requires more proposals because of its inefficient sampling of $\exp[-A(\mathbf{X})]$. In addition to the above analyses of the $O((DM)^{5/4})$ rule and the actual computation time, the Precision Annealing HMC method could achieve a significant speedup if run in parallel. We will again use Lorenz96 to illustrate this. The only change in Eq. (7.31) for a different system would be the choice for the vector field $f_a(\mathbf{x}(m), \boldsymbol{\theta})$. We will focus on the evaluation of $\nabla A(\mathbf{X})$ since this is the only calculation that scales directly with D and M.

- For the D vector field $\mathbf{F}_a(\mathbf{x}(m), v)$ in Eq. (7.18) and using Eq. (7.30) the time consuming part in calculating the D components in $\nabla A(\mathbf{X})$ at $t = t_m$ is to multiply the Jacobian with a D vector. This can be fully parallelized to realize a D times speedup.

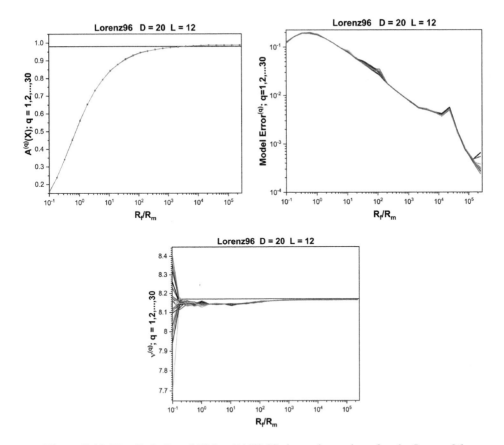

Figure 7.12 **Top Left Panel** Using PAHMC the action values for the Lorenz96 model with $D = 20$ and $L = 12$ are displayed. At each step in the precision annealing procedure corresponding to one R_f value, we perform Hamiltonian Monte Carlo calculations and come up with a path that is an average of all the paths proposed by HMC. This averaged path is used as the initial condition for the next R_f value. $N_I = 30$ independent calculations are performed at each R_f, resulting in the many action curves displayed. All the action levels quickly plateau at the anticipated level as R_f grows. **Top Right Panel** The model errors in Eq. (7.31) as a function of R_f for the Lorenz96 model with $D = 20$ and $L = 12$. $N_I = 30$ independent calculations are present. At smaller R_f values, the model error dominates the action. After that, the model error rapidly decreases as R_f grows. As a result, the measurement error term in Eq. (7.31) dominates the action and it becomes essentially independent of R_f. This indicates that the Precision Annealing procedure has successfully located the path that agrees well with the measurements and the model. **Bottom Panel** The estimations of the Lorenz96 model forcing parameter v at each value of R_f. $N_I = 30$ independent calculations are performed. The true forcing parameter in the twin experiments is $v = 8.17$ (solid black line).

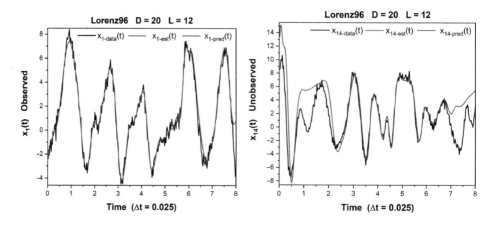

Figure 7.13 **Left Panel** Display of results using PAHMC for an *observed* dynamical variable $x_1(t)$ of the Lorenz96 model with $D = 20$, $L = 12$. The data, with noise added, span the full $0 \leq t \leq 11$ interval; they are displayed in black. The estimated values in the observation window $0 \leq t < 5$ are shown in red. The predicted values, with errors (in magenta), are shown in blue within the interval of $5 < t \leq 8$. **Right Panel** Display of results for an *unobserved* dynamical variable $x_{14}(t)$ of the Lorenz96 model with $D = 20$, $L = 10$. The true data, with noise added, span the full $0 \leq t \leq 8$ interval. The estimated values in the observation window $0 \leq t < 5$ are shown in red. The predicted values, with errors (in magenta), are shown in blue within the interval of $5 < t \leq 8$. For the unobserved variables, the data within $0 \leq t \leq 8$ are not presented to the PAHMC method.

- To complete the calculation of $\nabla A(\mathbf{X})$, we need to repeat the above process for each time point m from 1 to M. These repetitions are independent and can therefore be fully parallelized to accomplish an M times speedup.

If these two results are achieved with parallelization, then a scaling with D and M might be reduced to $O(DM^{1/4})$.

Clearly parallel computing for either Monte Carlo method is a good idea, and the results of Quinn and Abarbanel (2010); Quinn (2010) suggest that if the computational problems are large enough one can also expect to see a speedup proportional to the number of 'cuda cores' as well.

8

Machine Learning and Its Equivalence to Statistical Data Assimilation

8.1 $\langle A(\mathbf{X}) \rangle = \langle -\log P(\mathbf{X}|\mathbf{Y}) \rangle$; Action Is Information

Supervised Machine Learning is a framework in which information in some noisy *input* data is presented to a layer called l_0. This input is sampled M times, carrying information about the probability distribution of the input. We label this input $\mathbf{y}^{(k)}(l_0)$; $k = 1, 2, \ldots, M$. The input is transferred to and through a layered model network following a set of rules for the transfer from layer to layer. The aim is to produce at the final *output* layer, l_F, labeled, known characteristics of those data.

In such a task we could imagine as input at l_0 M sample images containing cats, schoolbuses, and roads: $\mathbf{y}^{(k)}(l_0)$; $k = 1, 2, \ldots, M$. The dynamical degrees of freedom of the model network $\mathbf{x}^{(k)}(l)$ are D-dimensional vectors at each layer l_j; $\{j = 1, \ldots, F-1\}$ evolving as we move from layer to layer during a training window $[l_0, l_F]$. We ask the machine to 'learn,' via some training protocol, to tell us at the output $l = l_F$ $\mathbf{x}^{(k)}(l_F)$ how many of each category (cats, schoolbuses, and roads) are present in the new images. The supervision is often human involvement categorizing the outputs.

Movement from layer to layer is determined by a rule $\mathbf{x}^{(k)}(l+1) = \mathbf{f}(\mathbf{x}^{(k)}(l), \boldsymbol{\theta}(l))$ where $\boldsymbol{\theta}(l)$ are a collection of $N_p(l)$ parameters at each layer l shaping the nonlinear function $\mathbf{f}(\mathbf{x}^{(k)}(l), \boldsymbol{\theta}(l))$.

From the information contained in the $k = 1, 2, \ldots, M$ inputs to the network model and their associated outputs from the network model, we seek to estimate the response of the $\mathbf{x}(l \neq \{l_0, l_F\})$ and the $N_p(l)$ parameters $\boldsymbol{\theta}(l)$ in the model. Once the estimation is completed and the network is trained, we may present additional inputs from the same source and ask how well the model with trained $\boldsymbol{\theta}(l)$ predicts correct new outputs $\mathbf{y}^{(k)}(l_F)$. The latter is called generalization and serves to validate the model as trained.

The difference between this setup and a similar description of data assimilation is threefold:

- the model parameters $\theta(l)$ depend on the layer,
- one has no rule linking $\theta(l+1)$ and $\theta(l)$, so the number of parameters might well seem huge. In the SDA context this indicates an 'external force' and yields constraints on the layer to layer dynamics, and
- the "hidden" steps in data assimilation are taken by moving the dynamical model from observation time τ_k to the next observation time τ_{k+1} in units of Δt. The operation is much the same in ML as SDA, but nothing is 'hidden' in the latter.

There seems to be no reason why the number of input or output locations in a supervised network could not be more than just one at the input and one at the output. There would then be many supervisors and many inputters doing their jobs at different layer locations.

Parameters such as these $\theta(l)$ are encountered in control theory, formulated in the 1930s and explained in many books, including Pontryagin (1959); Gelfand and Fomin (1963); Kirk (1970). They result in constraint equations in a variational formulation.

We also discuss a framework in which the time or layer label is taken to be continuous, providing a differential equation, the Euler-Lagrange equation and its boundary conditions, as a necessary condition for a minimum of the cost function or action. This shows that the problem being solved is a two point boundary value problem familiar in the discussion of variational methods (Gelfand and Fomin (1963)).

We call the use of continuous layers "deepest learning." In the ML literature, adding many layers to provide a 'representation' of the information presented to the network is often called deep learning (Goodfellow et al. (2016)). Making the layer variable continuous is as 'deep' as one can go. These problems respect a symplectic symmetry (Gelfand and Fomin (1963); Kot (2014); Arnol'd (1989)) in continuous time/layer phase space. Both Lagrangian versions and Hamiltonian versions of these problems are presented. When it comes to numerical evaluation of expected values in the MLP, we, once again, are required to use discrete layer steps. Maintaining the symplectic symmetry of the problem will turn to using our earlier discussions of this.

8.2 General Discussion of ML

8.2.1 Machine Learning; Standard Feedforward Neural Nets

We begin with a brief characterization of simple architectures in feedforward neural networks (Mitre17 (2017); Goodfellow et al. (2016); LeCun et al. (2015)). A graphical guide to this development is in Fig. 8.1.

The network is composed of an input layer l_0 and an output layer l_F and hidden layers $l_1, l_2, \ldots, l_F - 1$.

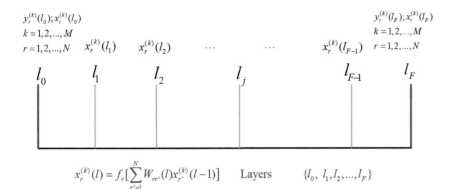

$$x_r^{(k)}(l) = f_r[\sum_{r'=1}^{N} W_{rr'}(l)x_{r'}^{(k)}(l-1)] \qquad \text{Layers} \qquad \{l_0,\ l_1, l_2, ..., l_F\}$$

*Data is presented at $l_0 : y(l_0)$—the **input** and at $l_F : y(l_F)$—the **output***

Figure 8.1 A graphic indicating the ingredients of a feed forward layered network $[l_0, l_F]$ with F+1 layers. An input layer $l = l_0$ receives M noisy inputs $\mathbf{y}^{(k)}(l_0)$; $k = 1, 2, \ldots M$ and is instructed at an output layer $l = l_F$ what the correct output $\mathbf{y}^{(k)}(l_F)$ should be. Imposing this supervision via the minimization of an action trains the network: that is, estimates the weights $\mathbf{W}_{rr'}(l)$ and any other parameters $\boldsymbol{\theta}(l)$ in $\mathbf{f}(\mathbf{Wx}^{(k)})$. The dimension of the active nonlinear units $\mathbf{x}^{(k)}(l)$ could be different at each layer: $l = \{l_0, l_2, \ldots, l_F\}$. As the layers $\{l_1, l_2, \ldots, l_{F-1}\}$ neither receive data input nor are they queried about their output, they are called *hidden layers*. When each layer l receives instructions from layer $l - 1$, this is known as a multi-layer perceptron (MLP). Once trained, the parameters along with the network architecture will take in new inputs and produce the correct outputs with some level of fidelity. Each input and output is a vector $y_r^{(k)}(l_0 \text{ or } l_F)$; $r = 1, 2, \ldots N_0$ or $r = 1, 2, \ldots, N_F$, and they need not be the same dimension.

Information is available to the network at l_0 and l_F. These layers are, in that regard, similar to the time points τ_k where data (and the information contained therein) is received in data assimilation.

Within each layer we have N_l active units, called 'neurons.' For our purposes the 'neurons' (not to be at all confused for the neurons the reader brings to the task of reading this book) in each layer are assigned the same role and structure as a nonlinear map in a dynamical system. This can be generalized to different numbers and different types of neurons in each layer at the cost of a notation explosion.

Data is available at layer l_0 and at layer l_F in many pairs of N_0-dimensional input and N_F-dimensional output. For simplicity of discussion, we choose $N_0 = N_F = N$. Then these data are pairs of N-dimensional vectors: $\{y_r^{(k)}(l_0), y_r^{(k)}(l_F)\}$; $r = 1, 2, \ldots, N$ where $k = 1, 2, \ldots M$ identifies the pairs selected from the data set we have at hand.

In the training of the network we will present M pairs of input/output, and this plays the role of the number of measurements at each observation time in the SDA formulation: it provides information flowing into and through the network architecture. When the network has learned the information available within the data, one might expect the action $\langle A(\mathbf{X}) \rangle = \langle - \log P(\mathbf{X}|\mathbf{Y}) \rangle$, which we see is related to the entropy, to saturate as a function of M

The activity of the units in each 'hidden' layer l, $x_r^{(k)}(l)$; $l_0 < l < l_F$ is determined by the activity in the previous layer. This rule for moving from layer $l - 1 \to l$ is described by a nonlinear function; we chose this one:

$$x_r^{(k)}(l) = f_r(\mathbf{x}^{(k)}(l-1), l) = f_r \left(\sum_{r'=1}^{N} W_{r\,r'}(l) x_{r'}^{(k)}(l-1) \right). \qquad (8.1)$$

In addition to any other parameters $\boldsymbol{\theta}(l)$ in the nonlinear function $f_r(\bullet)$, the weights $W_{rq}(l)$ must be estimated. The summation over weights $W_{rq}(l)$ determines how the activities in layer $l - 1$ are combined before allowing $f_r(\mathbf{x}(l), \boldsymbol{\theta}(l))$ to act, yielding the activities at layer $l + 1$. There are numerous choices for the manner in which the weight functions act as well as numerous choices for the nonlinear functions, and we direct the reader to the references for the discussion of the virtues of the various choices (Mitre17 (2017); Goodfellow et al. (2016); LeCun et al. (2015)).

At the input and the output layers the network activities $\{\mathbf{x}^{(k)}(l_0), \mathbf{x}^{(k)}(l_F)\}$ are compared to M pairs of input/output noisy data $\{\mathbf{y}^{(k)}(l_0), \mathbf{y}^{(k)}(l_F)\}$, and the network performance is assessed using an error metric, often a least squares criterion, or cost function like this:

$$C_M(\mathbf{x}^{(k)}(l), \mathbf{y}^{(k)}(l)) = \frac{1}{M} \sum_{k=1}^{M} \frac{1}{2N} \sum_{r=1}^{N} R_m(r, l) \left[x_r^{(k)}(l) - y_r^{(k)}(l) \right]^2, \qquad (8.2)$$

where $R_m(r, l)$ is nonzero only at $l = \{l_0, l_F\}$. Minimization of this cost function over all activity at each layer $x_r^{(k)}(l)$ and weights $W_{rq}(l)$, **subject to** the network model Eq. (8.1), is used to determine the weights, the variables $x_r^{(k)}(l)$ in all layers, and any other parameters $\boldsymbol{\theta}(l)$ appearing in the architecture of the network.

The matrices $\mathbf{W}(l)$ defining the architecture of the network are taken independent of the index k indicating the number of input samples $k = 1, 2, \ldots, M$ as they are part of the network specification.

When we are presenting $k = 1, 2, \ldots, M$ pairs of input/output data, the activities $\mathbf{x}^{(k)}(l)$ at each node of the network will vary, as each noisy input data item will be drawn from a distribution of inputs, but the overall network architecture is dictated by the weights $\mathbf{W}(l)$, which we learn from the response to this distributions of inputs.

We wish to find the global minimum of the cost function Eq. (8.2), **subject to** the network model Eq. (8.1), which is a nonlinear function of the neuron activities as well as the weights and any other parameters in the functions at each layer. This is an NP-complete problem as discussed previously.

8.2.2 Machine Learning with Model Error

The ML problem as described here (Mitre17 (2017); Goodfellow et al. (2016)) assumes there is **no error** in the model Eq. (8.1). This is a bold statement of knowledge about layer-to-layer network rules. This deterministic statement results in the outputs at layer l_F, $\mathbf{x}^{(k)}(l_F)$, being very complicated functions of the parameters in the model and the activities at layers $l \leq l_F$. We relax the equality constraint Eq. (8.1) by adding it as a 'penalty function' (Kirk (1970)) to the cost function, defining the ML 'action' $A_{ML}(\mathbf{X})$:

$$
\begin{aligned}
A_{ML}(\mathbf{X}) = \frac{1}{M} \sum_{k=1}^{M} &\left\{ \frac{1}{2N} \sum_{r=1}^{N} R_m(r, l_0) \left[x_r^{(k)}(l_0) - y_r^{(k)}(l_0) \right]^2 \right. \\
&\left. + \frac{1}{2N} \sum_{r=1}^{N} R_m(r, l_F) \left[x_r^{(k)}(l_F) - y_r^{(k)}(l_F) \right]^2 \right\} \\
&+ \frac{R_f}{2} \sum_{l=l_0}^{l_F-1} \left[\sum_{j=1}^{N} x_j^{(k)}(l+1) - f_j\left(\sum_{i=1}^{N} W_{j,i}(l) x_i^{(k)}(l) \right) \right]^2.
\end{aligned}
\quad (8.3)
$$

In the limit $R_f \to \infty$ the equality constraint on the layer-to-layer rule is restored. Another viewpoint casts the layer-to-layer rule as stochastic with additive Gaussian noise and a diagonal precision matrix R_f:

$$
x_j^{(k)}(l+1) = f_j\left(\sum_{i=1}^{N} W_{j,i}(l) x_i^{(k)}(l) \right) + \frac{1}{\sqrt{R_f}} \mathcal{N}(0, 1).
\quad (8.4)
$$

We interpret the action as $A_{ML}(\mathbf{X}) = -\log[P(\mathbf{X}|\{y_r^{(k)}(l_0), y_r^{(k)}(l_F)\})]$.

8.3 Using ML to Predict Subsequent Terms in a Time Series {s(n)}

We are given a long time series of scalar data $s(n)$; $n = 1, 2, \ldots$, and asked to build an ML model that will receive as input a sequence of that time series, say $\{s(k), s(k+1), \ldots, s(k+n)\}$ and as output produce the next element $\{s(k+n+1)\}$ of the given time series.

Using standard nonlinear time series analysis (Abarbanel (1996); Kantz and Schreiber (2004)) we find the dimension of the attractor for the time series is about 4.4. The system from which the scalar time series came has two positive Lyapunov exponents and one zero exponent (Oseledec (1968)). The latter item tells us the time series arises from a continuous time dynamical system. We also determine that the scalar signal comes from a projection of a $D_E = 5$ dimensional system (Ty et al. (2019)). The idea is now to work in $D_E = 5$ dimensional space on vectors $\mathbf{s}(n) = [s(n), s(n + \tau), s(n + 2\tau), \ldots, s(n + (D_E - 1)\tau)]$, and build a machine that learns the discrete time mapping $\mathbf{s}(n) \to \mathbf{s}(n + 1)$.

To the scalar data $\mathbf{s}(n)$ we add noise of mean zero and rms error σ to form noisy scalar data $y(n) = s(n) + \text{Noise}(0, \sigma)$. We selected σ to be 2% of the dynamical range of the observed data.

Using these noisy scalar data, we form a data library of many input/output D_E-dimensional vector pairs to be used at the input port at layer l_0 and the output port at layer l_F to train a neural network. We use $k = 1, 2, \ldots, M$ members of this library as our training set.

$$
\begin{aligned}
\mathbf{Y}^{(k)}(l_0) &= \{y(k), y(k + \tau), y(k + 2\tau), y(k + 3\tau), y(k + 4\tau)\}, \\
\mathbf{Y}^{(k)}(l_F) &= \{y(k + 1), y(k + 1 + \tau), y(k + 1 + 2\tau), \\
&\qquad y(k + 1 + 3\tau), y(k + 1 + 4\tau)\},
\end{aligned}
\tag{8.5}
$$

$k = 1, 2, \ldots, M$ and $D_E = 5$.

The network we selected is a Multi-layer Perceptron, and we wish to train it to receive the D_E-dimensional input vectors $\mathbf{Y}^{(k)}(l_0)$ and produce at the output D_E-dimensional vectors $\mathbf{Y}^{(k)}(l_F)$. At the input layer l_0 we have **one** input port with D_E slots. At the output layer l_F we have **one** output port with D_E slots. The network has $l_F - 2$ hidden layers $l = \{l_1, l_2, \ldots, l_F - 1\}$. At the hidden layers we have D_{hl} active units ('neurons') at layer l. As a function of the three quantities $\{l_F, D_{hl}, M\}$: l_F, the number of layers or the 'depth' of the network; D_{hl}, the number of active units in layer l or the 'breadth' of the network; and M, the number of distinct input/output pairs containing the information presented to the network for training, we wish to analyze the quality of the training, the accuracy of the operation of the trained network on input/output pairs **not** used in training, and the ability of the trained network to represent the information in the M data pairs. In the networks we develop here, we take D_{hl} to be independent of l. This not necessary; it only serves to simplify the task (and the notation). We use $D_{h0} = D_{hF} = D_E = 5$.

8.3.1 The Action

In much of machine learning one seeks to minimize a cost function evaluated at the input (l_0) and the output (l_F) layers of a selected network. We call the active variables ('neurons') at layer l $x_q^{(k)}(l)$ for active unit $q = 1, 2, \ldots, D_{hl}$ in layer l.

The measurement cost function for each input/output pair is at time k

$$C(k, D_{h0}, D_{hF}) = \frac{R_m}{2} \frac{1}{D_{h0} + D_{hf}} \left[\sum_{q=1}^{D_{h0}} (x_q^{(k)}(l_0) - y(k + (q-1)\tau))^2 \right.$$

$$\left. + \sum_{q=1}^{D_{hF}} (x_q^{(k)}(l_F) - y(k + 1 + (q-1)\tau))^2 \right], \qquad (8.6)$$

where the noise or errors in the input and output data have been taken to be Gaussian with zero mean and diagonal precision matrix R_m.

This is to be minimized subject to a layer-to-layer connection rule

$$x_q(l+1) = f_q\left(\sum_{v=1}^{D_{hl}} \mathbf{W}(l)_{qv} x_v(l)\right) \quad q = 1, \dots D_{h(l+1)}, \qquad (8.7)$$

with $\mathbf{W}(l)$ a matrix of weights to be determined in the minimization of $C(k)$.

If Gaussian errors with precision matrix R_f are accepted in the layer-to-layer rule Eq. (8.7), then the action is

$$A(x(l); k) = C(k, D_{h0}, D_{hF}) +$$

$$\frac{1}{l_F - l_0} \sum_{l=l_0}^{l=l_F} \frac{R_f}{2} \sum_{q=1}^{D_{h(l+1)}} \left(x_q^{(k)}(l+1) - f_q\left(\sum_{v=1}^{D_{hl}} \mathbf{W}(l)_{qv} x_v^{(k)}(l)\right)\right)^2 \qquad (8.8)$$

and we call this the 'action,' after its usage in statistical Physics, for a single input/output data pair chosen at time k. When we have many input/output pairs, we add a label to the active states in the network $x_q(l) \rightarrow x_q^{(k)}(l)$, and our goal is to minimize the action

$$A(x_q^{(k)}(l), \ \mathbf{W}(l)) = \frac{1}{M} \sum_{k=1}^{M} \left\{ \frac{R_m}{2} \frac{1}{D_{h0} + D_{hf}} \left[\sum_{q=1}^{D_{h0}} (x_q^{(k)}(l_0) - y(k + (q-1)\tau))^2 \right. \right.$$

$$\left. + \sum_{q=1}^{D_{hF}} \left(x_q^{(k)}(l_F) - y(k + 1 + (q-1)\tau)\right)^2 \right]$$

$$\left. + \frac{R_f}{2} \frac{1}{\sum_{l=l_1}^{l_F} D_{hl}} \sum_{l=l_0}^{l_F} \sum_{q=1}^{D_{h(l+1)}} \left(x_q^{(k)}(l+1) - f_q\left[\sum_{v=1}^{D_{hl}} \mathbf{W}(l)_{qv} x_v^{(k)}(l)\right]\right)^2 \right\},$$

$$(8.9)$$

with respect to the connection weight matrices $\mathbf{W}(l)$ and the activities $x_q^{(k)}(l)$. Minimizing this action recognizes that for each input/output training pair, the activity of the network nodes may differ, but taking into account the variation over all M

presentations of pairs from the library will train a possible generalizable network characterized by the $\mathbf{W}(l)$ and any other fixed parameters $\boldsymbol{\theta}(l)$ in the nonlinear functions $f_q(\mathbf{x}(l), \boldsymbol{\theta}(l))$.

8.4 Action Levels for the Given Time Series $\{s(n)\}$

Now we build and train a feedforward MLP (Rozdeba (2018)) to learn the function $s(n) \rightarrow s(n + 1)$ using D_E-dimensional data pairs from our library. We examined networks with $D_E = 5$ dimensional input (l_0) and output (l_F) layers and 1–5 hidden layers each with the same number D_h of active units ('neurons'). The nonlinear function operating from layer to layer was chosen to be tanh($\bullet$). We use the Python based program VarrAnneal Rozdeba (2018) to perform the minimization of the action at each value of $R_f/R_m > 0$. This is an instantiation of Precision Annealing, and we recall that it was discussed in Chapter 5. To prepare our data for these network choices, we first scaled all of the noisy inputs $y(n)$ to lie within the range $[-1, 1]$ via

$$y(n) \rightarrow \frac{2y(n) - (y_{max} + y_{min})}{y_{max} - y_{min}}, \tag{8.10}$$

where $y_{max,min}$ are the maximum and minimum values taken by the noisy data. These scaled values were used to construct our data library of input/output pairs.

8.4.1 Two Hidden Layers; $D_h = 15$; $M = 50, \ldots, 1200$

Using PA and systematically moving R_f/R_m from $R_{f0}/R_m \approx 10^{-8}$ to $R_f/R_m \approx 10^{11}$ we evaluated $A(\mathbf{X})$ for $M = 50, 100, \ldots, 1200$ input/output pairs. R_f was slowly increased, using $R_f/R_m = R_{f0}/R_m \alpha^\beta$ $\alpha = 1.1$. At each value of R_f/R_m we used $N_I = 20$ initializations of the PA procedure.

We first examine the structure of the action levels as a function of R_f/R_m for $M = 50, 300, 900, 1200$ I/O pairs. This is displayed in Fig. 8.2. Note that the action levels become nearly independent of R_f/R_m for large values of this hyperparameter. Equally interesting is the initial rise of $A(\mathbf{X})$ for large R_f/R_m as M increases. Then this saturates as the information in the time series $s(n)$ is represented fully in the network. See Fig. 8.3. Recall we call this quantity the information content as, up to a constant $\langle A(\mathbf{X})\rangle = \langle -\log[P(\mathbf{X})]\rangle$.

8.5 Errors in Training and Validation

Once we have trained the proposed network, we can evaluate its quality when performing the task we have set it. In the example we have discussed here, that task is summarized as: when presented with a D_E-dimensional vector of inputs, created

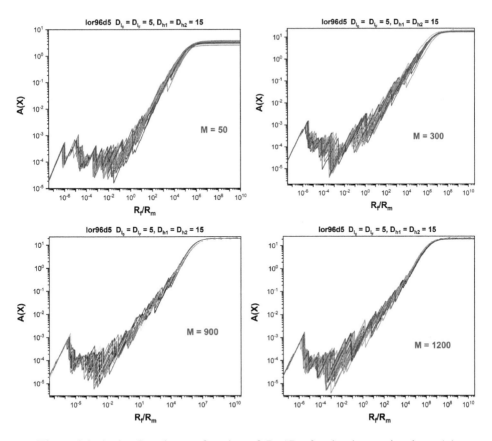

Figure 8.2 Action Levels as a function of R_f/R_m for the time series data $s(n)$ input into our two hidden layer MLP as a D_E-dimensional data vector. The number of I/O pairs for these calculations were $M = 50, 300, 900, 1200$. $N_I = 20$ action levels associated with the N_I initializations of the optimization algorithm used at each value of R_f/R_m. In these calculations $\alpha = 1.1$ in the Precision Annealing procedure. Note that as M increases, the action level for large R_f/R_m rises and then saturates, becoming effectively independent of R_f/R_m. This is presented as well in Fig. 8.4.

from time delays of a signal $s(n)$, accurately produce the next element of the time series $s(n + 1)$. The quantities $s(n)$ and $s(n + 1)$ are the first components of the data vectors.

We have tested (or validated) the operation of the network both on the data used to train the network and on data held aside in our library of I/O pairs. The latter is often called the "test" set or validation set or prediction set portion of the total data available to us (Frank et al. (2001)).

The error on the training set as a function of the number M of I/O pairs used to train the network is given as

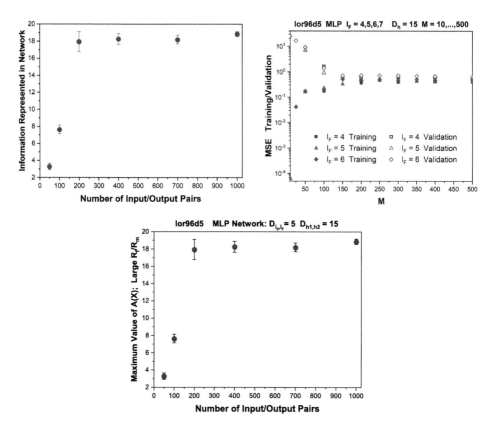

Figure 8.3 **Upper Left Panel** Information represented in the model network as a function of the number of samples M. **Upper Right Panel** The Mean Square Errors (MSEs) in training and validation (estimation and prediction) for the network architecture with $l_F = 4, 5, 6$ layers. The input and output layers have five ports as before. All hidden layers have 15 active units. We see that when $M \geq 200$ or so, the performance of the network architecture becomes independent of the number of training samples as well as of the number of layers in the network. This result shows how the PA method can capture essential information processing power of a selected architecture of a multi-layer perceptron. More to the point, it informs us how many (M) input/output pairs are required to perform the task set to the machine. **Bottom Panel** Display of the average and standard deviation of the $N_I = 20$ largest action levels versus the number M of I/O pairs in the training set for our time series $\{s(n)\}$. As seen in the action levels plots, as M increases, the maximum action levels grow then saturate as the network reaches a full representation of the information in the data pairs. This kind of calculation allows the network designer to determine for a given network architecture how many samples from the I/O library of pairs will be needed to fully train the network. The expected $A(\mathbf{X})$ saturates near $M = 200$.

$$MSE_{\text{Training Error}}(M) = \frac{1}{M}\sum_{k=1}^{M}\frac{1}{D_E}\sum_{q=1}^{D_E}(x_q^{(k)}(l_F) - y(k+1+(q-1)\tau))^2 \quad (8.11)$$

This compares, in a least squares sense, the D_E-dimensional output $x_q^{(k)}(l_F)$ from the trained network with the data from the **training** set, $y(k+1+(q-1)\tau)$, that are the output side of the input/output training pairs. The input to the trained network is the values $\mathbf{Y}^{(k)}(l_0) = \{y(k), y(k+\tau), y(k+2\tau), y(k+3\tau), y(k+4\tau)\}$; the trained network operates on this D_E-dimensional vector producing the output $x_q^{(k)}(l_F)$; $q = 1, 2, \ldots, D_E$. We plot this as a function of M, the number of input/output pairs used in the training procedure, and in Fig. 8.3 we also examine the dependence on the number of active units ("neurons") in each of the two hidden layers in the network.

We also can determine the accuracy of the trained network when acting on inputs selected from I/O pairs *not* used in the training of the network. This 'validation' error is evaluated as

$$MSE_{\text{Validation Error}}(M) =$$

$$\frac{1}{M_{\text{total}} - M}\sum_{k=M}^{M_{\text{total}}}\frac{1}{D_E}\sum_{q=1}^{D_E}\left(x_q^{(k)}(l_F) - y(k+1+(q-1)\tau)\right)^2. \quad (8.12)$$

This compares the D_E-dimensional output $x_q^{(k)}(l_F)$ from the trained network with the data from the set of input/output pairs that **were not used** during the training, $y(k+1+(q-1)\tau)$, that are the output side of the input/output pairs from the data library. We plot this as a function of M, the number of input/output pairs used in the training procedure. All of the I/O pairs from the data library not used in training were used in this validation error estimate. $M_{\text{Total}} = 84971 \gg M$.

The results in Fig. 8.4 show that for small M the training and validation errors differ substantially, but as M increases, enough information lies in the training set of M training I/O pairs that the overall training error levels out when the network has completed its representation of the information in the data. Similarly, while the validation error is large for small M, as the network becomes 'well trained' (represents the information in the data series) the prediction MSE is essentially the same as the MSE in training. This result is consistent with the observation that the maximum value of the action levels for large R_f/R_m becomes independent of M; see Fig. 8.3.

Fig. 8.3 examines the training and validation MSEs as a function of the number of layers in the network. The number of hidden layers is $l_F - 2$, and we have evaluated this, using our architecture, for $l_F = 4, 5,$ and 6.

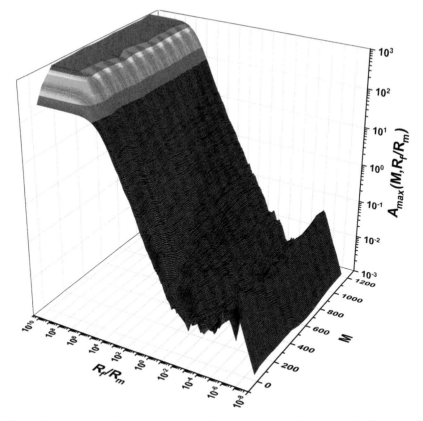

Figure 8.4 A three-dimensional plot of the action for the $l_F = 4$; $D_h = 15$ network as a function of R_f/R_m and the number of input/output pairs M presented for training. It is clear here that a distinct plateau appears in the action. This means that the information content $(-\log[P])$ within the time series data is now fully represented in the network.

We display the dependence of the action on the number of input/output samples M and R_f/R_m relevant in the PA algorithm in Fig. 8.3 and Fig. 8.4 to further illustrate the outcome of our MLP network instantiation.

In Fig. 8.5 we display the predictions produced by our trained MLP network after the training using $M = 400$ input/output pairs of segments of the noisy time series starting at $y(n-1)$ and predicting $y(n)$ in comparison with the known value of $y_{data}(n)$.

The training is performed in D_E-dimensional space, $D_E = 5$, and the output of the network is also in D_E-dimensional space. We display only the first component of the $D_E = 5$ dimensional proxy state space vector, as that is our (noisy) measured quantity. The predictions are what this network has been trained to do.

If we ask another question of the network: take the trained network as a dynamical system, namely, train the network using $M = 400$ input/output pairs,

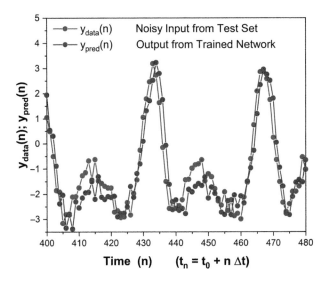

Figure 8.5 Using the $l_F = 4$; $D_{in} = D_{out} = 5$; $D_h = 15$ network, we show how, after training with M = 400 pairs of $D_E = 5$ dimensional noisy inputs $\{s(n), \ldots, s(n + 5)\}$ and 5 dimensional outputs $\{s(n + 1), \ldots, s(n + 6)\}$, this network is able to predict one step ahead for a new five-dimensional input.

then use the trained network to predict $y(n) \rightarrow y(n + 1)$ forward from the training window, we find that on this task the trained network does not perform as well as on the task it was trained to do

8.6 "Twin Experiment" with a Multi-layer Perceptron

8.6.1 Creating Data with a Known Network

We built a feedforward network with l_F layers: one input layer at l_0 and one output layer at l_F. This network has $l_F - 2$ hidden layers. A network with F+1 = 100 layers was constructed with weights drawn from a uniform distribution $U[-0.1, 0.1]$. The network has 10 neurons in each layer. The activity in neuron j in layer l, $x_j(l)$, is related to $x_j(l - 1)$ in the previous layer as

$$x_j(l) = g(\mathbf{W}(l)\mathbf{x}(l - 1)); \quad g(z) = 0.5\left[1 + \tanh\left(\frac{z}{2}\right)\right]. \tag{8.13}$$

Data, consisting of input/output pairs $\{x_i^{(k)}(l_0), x_i^{(k)}(l_F)\}$; $k = 1, 2, \ldots$, were constructed by presenting $x_i^{(k)}(l_0)$ at layer l_0 and passing it through the network to generate $x_i^{(k)}(l_F)$ at the output layer l_F. To these input/output pairs we added Gaussian noise with zero mean and variance 0.0025 to create our noisy library of

$\{\mathbf{y}^{(k)}(l_0), \mathbf{y}^{(k)}(l_F)\}; \quad k = 1, 2, \ldots, M$ pairs. These noisy data were now stored for further use. This prepares us for a supervised learning task.

8.6.2 Selecting a Model to Represent the $\{\mathbf{y}^{(k)}(l_0), \mathbf{y}^{(k)}(l_F)\}$ Pairs

We are now presented with a subset of this library of input/output pairs and choose a multi-layer perceptron with l_F layers and N neurons per layer to represent the input/output relationship: $\{y_i^{(k)}(l_0), y_i^{(k)}(l_F)\}; \quad k = 1, 2, \ldots, M$. This selected network had $N = 10$ neurons in each of $l_F = 10$ layers, at first. At the input and output layers, ten noisy values $\{y_i^{(k)}(l_0), y_i^{(k)}(l_F)\}; \quad i = 1, \ldots, N = 10$ are presented for various values of M. This corresponds to $L = 10$ in the notation used before. As M is increased, more and more of the complexity of the relationship of the pairs is explored and more and more is demanded of our selected network.

We purposely chose to explore the relationship of our noisy data set pairs $\{\mathbf{y}^{(k)}(l_0), \mathbf{y}^{(k)}(l_F)\}; \quad \mathbf{y} = \{y_1, y_2, \ldots, y_L\}$ with a **wrong** model. Our goal is to see where the incorrectness of the model displays itself, and then to augment the original model choice to improve its ability to predict outputs from inputs it had not seen before.

M input/output pairs are presented to the model with 10 inputs $\mathbf{y}^{(k)}(l_0)$ at layer l_0 and 10 data outputs $\mathbf{y}^{(k)}(l_F)$ at layer l_F. We investigated presenting $k = 1, 2, \ldots, M$ samples of input/output pairs, choosing $M = 1, 2, 10,$ and 100.

In each case we minimized the action over all the weights and all the states $x_a^{(k)}(l)$ at all layers of the model for all $k = 1, 2, \ldots, M$ presentations of the samples of the data:

$$A_{ML}(\mathbf{X}) = \frac{1}{M} \sum_{k=1}^{M} \left\{ \frac{R_m}{2L} \sum_{i=1}^{N} \left[(x_i^{(k)}(l_0) - y_i^{(k)}(l_0))^2 + (x_i^{(k)}(l_F) - y_i^{(k)}(l_F))^2 \right] \right.$$

$$\left. + \frac{R_f}{Nl_F} \sum_{l=l_0}^{l_F-1} \sum_{a=1}^{N} \left[x_a^{(k)}(l+1) - g_a(\mathbf{W}(l)\mathbf{x}^{(k)}(l)) \right]^2 \right\}. \tag{8.14}$$

We use the Precision Annealing procedure described in Chapter 5 to identify the action levels for various paths through the network. The initial value of R_{f0}/R_m is taken to be 10^{-8}, and this is incremented via $R_f/R_m = R_{f0}\alpha^\beta$ with $\alpha = 1.1$ and $\beta = 0, 1, \ldots,$. In the numerical optimizations for the ML example we used L-BFGS-B (Byrd et al. (1995); Zhu et al. (1997)).

To present more information to our selected model we could increase the number of training pairs available to the network at l_0 and l_F; this is our number M. M can be chosen as large as the user wishes. To augment the ability of the model to represent the complexity of the data set as M increases, we also could increase N or l_F or both.

We anticipate that increasing the number of training pairs M will lead to a saturation of the action $A_{ML}(\mathbf{X})$ as a function of R_f/R_m when this hyperparameter grows large indicating that the number of input/output pairs where this occurs is where the model has captured all the information available in the data set we were presented. This provides a practical algorithmic tool to decide when we need no more data from our noisy data library.

We did not explore the importance of presenting fewer than N data elements at the input or output layer; $L < N$. We have done that for other examples not reported here, and as one increases L, the ability of the model to represent the data does improve. We also do not report here on using other nonlinear 'neuron' functions such as $g(x) = \log(1 + e^x)$, a ReLU like nonlinearity (Goodfellow et al. (2016); LeCun et al. (2015)). The results are not substantially different than what we have just seen.

Prediction with the Selected Model

Once we have used the PA procedure to select the path $\mathbf{X}$, comprised of the $x_j^{(k)}(l)$ and the $W_{ij}(l)$ for all neurons and weights at each layer $l_0 \leq l \leq l_F$, giving the minimum action $A(\mathbf{X})$, we have a new model that can be used to predict the output from presenting a new input. From the original data set of noisy pairs $\{y_j(l_0), y_i(l_F)\}$ we now select M_P unused pairs for testing the model. The inputs $y_j^{(r)}(l_0)$; $r = 1, 2, \ldots, M_P$; $j = 1, \ldots, N$ are presented to the input layer of the estimated model. The outputs from the operation of the model network $x_j^{(r)}(l_F)$ are compared to the known outputs $y_j^{(r)}(l_F)$ using the averaged square error

$$E(l_F)^2 = \frac{1}{NM_P} \sum_{r=1}^{M_P} \sum_{j=1}^{N} \left(x_j^{(r)}(l_F) - y_j^{(r)}(l_F) \right)^2. \tag{8.15}$$

Models to Represent the Pairs $\{\mathbf{y}^{(k)}(l_0), \mathbf{y}^{(k)}(l_F)\}$

All models have 10 neurons in each layer, and we examine different values of l_F and M, the number of training input/output pairs. In each calculation we used $N_I = 100$ initial paths in the PA procedure. In each graphic, therefore, there are 100 action levels possible. As the number of training pairs is increased, the number of action levels decreases as a function of R_f/R_m, so many, and then all, of the $N_I = 100$ initial conditions go to the same minimum of the action.

The PA procedure allows one to select a model among the choices one makes in the model design to accomplish the quality of prediction error desired. In the example data set we created and used here, many fewer layers were required than the number used ($l_F = 100$) to create the data set, and with the presentation of quite a small number of input/output pairs we could achieve small prediction errors. It

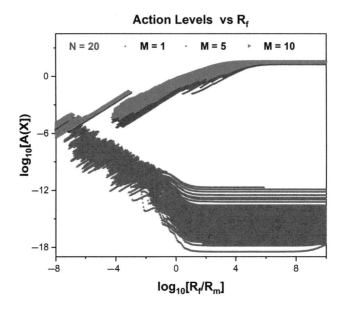

Figure 8.6 Action levels from training, estimating parameters (weights) and unobserved state variables, in a network with $l_F = 50$ layers and $N = 10$ neurons per layer. The data input/output pairs were generated with $l_F = 100, N = 10$ and chosen parameters. In each calculation we display 100 action levels at each value of R_f/R_m in the Precision Annealing protocol described in the text. The action levels (Eq. (8.14)) are shown as we increase R_f/R_m from a quite small value $R_{f0}/R_m = 10^{-8}$ to large values $R_f/R_m = (R_{f0}/R_m)\alpha^{\beta}$ with $\alpha = 1.1$ and $\beta = 0, 1, \ldots$, until $R_f/R_m \approx 10^{10}$. We display action levels for $M = 1, 2$, and 10 input/output pairs presented to the model network.

may go without saying that while we lack tools to simply 'examine' a set of data pairs and know which model (l_F, N) and architecture one would require and how many training pairs would also be required, PA provides a constructive path to addressing the question, based on the prediction accuracy the user desires.

The results of the procedures described here are found in Fig. 8.6, 8.7, and 8.8.

8.6.3 Recurrent Networks

In a recurrent network architecture one allows both interactions among neurons from one layer to another layer as well as interactions among neurons within a single layer (Jordan (1986); Elman (1990)). The activity $x_j(l)$ of neuron $j, j = 1, 2, \ldots, N$ in layer $l \{l_0, l_1, \ldots, l_F\}$ is given by $x_j(l) = f[\sum_i w_{ji}(l)x_i(l-1)]$ in a feedforward, `layer goes to the next layer`, network. We can add interactions within a layer in the same fashion, and to give some 'dynamics' to this within-layer activity we introduce a sequential label σ to the activity of neuron

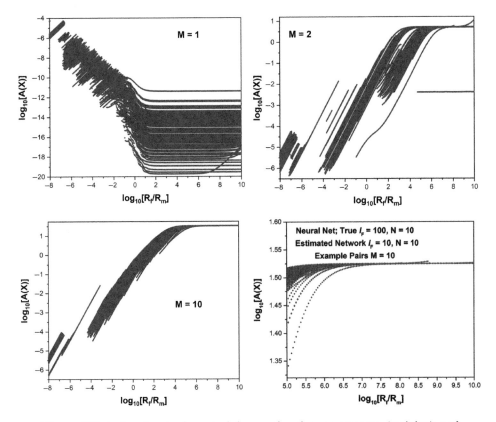

Figure 8.7 Action levels from training, estimating parameters (weights) and unobserved state variables, in a network with $l_F = 10$ layers and $N = 10$ neurons per layer. The data input/output pairs were generated with $l_F = 100$, $N = 10$ and chosen parameters. In each calculation we display 100 action levels at each value of R_f/R_m in the Precision Annealing protocol described in the text. The action levels (Eq. (8.14)) are displayed when we increase R_f/R_m from a quite small value $R_{f0}/R_m = 10^{-8}$ to large values $R_f/R_m = (R_{f0}/R_m)\alpha^\beta$ with $\alpha = 1.1$ and $\beta = 0, 1, \ldots$, until $R_f/R_m \approx 10^{10}$. **Upper Left Panel** $M = 1$ data pair. There is not enough information at $M = 1$ to produce an action level that dominates the integral Eq. (3.1). **Upper Right Panel** $M = 2$ data pairs. **Lower Left Panel** $M = 10$ data pairs. At large R_f/R_m only one action level remains. **Lower Right Panel** Blowup of the Lower Left Panel to show that only one action level remains for large enough R_f/R_m. The input/output pairs constituting the data were generated by a network with $N = 10$ neurons in each of $l_F = 100$ layers.

j in layer l: $x_j(l, \sigma)$. The mapping from layer to layer and within a layer can be summarized by

$$x_j(l, \sigma) = f\left[\sum_i W_{ji}(l) x_i(l-1, \sigma) + \sum_i w_{ji}(l) x_j(l, \sigma - 1) \right]. \qquad (8.16)$$

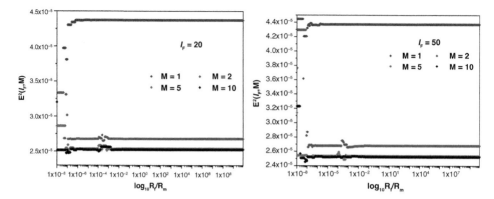

Figure 8.8 Prediction Errors Eq. (8.15) for proposed networks with $l_F = 20, 50$ layers and $M = 1, 2, 5, 10$ training pairs. In each case the path, hidden layer neuron activities and weights, producing the lowest minimum in the action levels determined the trained network. The number of 'test' pairs presented to the trained network was $M_P = 100$. The training data input/output pairs as well as the testing input/output pairs were generated from a network with $l_F = 100$ layers.

Another version of this allows the nonlinear function to be different for layer-to-layer connections and within-layer connections, so

$$x_j(l, \sigma) = f\left[\sum_i W_{ji}(l)x_i(l - 1, \sigma)\right] + g\left[\sum_i w_{ji}(l)x_j(l, \sigma - 1)\right], \quad (8.17)$$

where $f(x)$ and $g(x)$ can be different nonlinear functions.

We can translate these expressions into the DA structure by recognizing that $x_j(l)$ is the 'model variable' in the layer-to-layer function while in the recurrent network, the state variables become $x_j(l, \sigma)$. It seems natural that as dimensions of connectivity are added – here going from solely feedforward to that plus within-layer connections – that additional independent variables would be aspects of the 'neuron' state variables' representation.

In adding connections among the neurons within a layer we have another independent variable; we called it σ, and the 'point' neurons depending on layer alone become fields $x_j(l, \sigma)$. In the ML/AI networks we have no restrictions on the number of independent variables. This leads to the investigation of 'neural' fields $\phi_j(\mathbf{v})$ where $\mathbf{v}$ is a collection of independent variables indicating which layers are involved in the progression of the field from an input to an output layer.

However many independent variables and however many 'neurons' we utilize in the architecture of our model network, the overall goal of identifying the conditional probability distribution $P(\mathbf{X}|\mathbf{Y})$ and estimating the moments or expected values of interest still comes to one form or another of approximating integrals such as Eq. (3.1).

8.7 Continuous Layers: Deepest Learning

8.7.1 Euler-Lagrange Equations for ML; Lagrangian Formulation

Often one may discover structure when $\Delta l \rightarrow 0$ and the difference equations describing movement from layer to layer become differential equations, the Euler-Lagrange equations (Goldstein et al. (2002); Arnol'd (1989); Gelfand and Fomin (1963)). This was discussed in Chapter 6, and only the measurement error term in the action differs here. The limit of the action where the number of layers becomes a continuous variable is

$$A_0(\mathbf{x}(l), \mathbf{x}'(l)) = \int_{l_0}^{l_F} dl \, L(\mathbf{x}(l), \mathbf{x}'(l), l), \qquad (8.18)$$

where

$$
\begin{aligned}
L(\mathbf{x}(l), \dot{\mathbf{x}}(l), l) &= \sum_{r=1}^{N} \frac{R_m(r, l)}{2} \Big(x_r(l) - y_r(l) \Big)^2 + \\
&\quad \sum_{a=1}^{D} \frac{R_f(a)}{2} \left(\frac{dx_a(l)}{dl} - F_a(\mathbf{x}(l), \boldsymbol{\theta}(l)) \right)^2 \\
&\quad + \frac{1}{2} \nabla_l \cdot \mathbf{F}(\mathbf{x}(l), \boldsymbol{\theta}(l)) \\
&= \chi(\mathbf{x}(l) - \mathbf{y}(l)) + \\
&\quad \sum_{a=1}^{D} \frac{R_f(a)}{2} \left(\frac{dx_a(l)}{dl} - F_a(\mathbf{x}(l)) \right)^2 \\
&\quad + \frac{1}{2} \nabla_l \cdot \mathbf{F}(\mathbf{x}(l), \boldsymbol{\theta}(l)).
\end{aligned}
\qquad (8.19)
$$

$R_m(r, l)$ is nonzero only at $l = l_0$ and $l = l_F$.

Within the ML context we call this 'deepest learning' as the number of layers goes to infinity in an instructive manner.

The minimization of the action now requires that the paths $\mathbf{x}(l)$ in $\{\mathbf{x}(l), \mathbf{x}'(l)\}$ space satisfy the Euler-Lagrange equation $\frac{d}{dl} \left[\frac{\partial L(\mathbf{X}(l), \mathbf{X}'(l), l)}{\partial x_a'(l)} \right] = \frac{\partial L(\mathbf{X}(l), \mathbf{X}'(l), l)}{\partial x_a(l)}$. The solutions must also satisfy the boundary conditions $\sum_{a=1}^{D} \delta x_a(l_0) p_a(l_0) = 0$, $\sum_{a=1}^{D} \delta x_a(l_F) p_a(l_F) = 0$ where $p_a(l) = \partial L(\mathbf{x}(l), \mathbf{x}'(l), l)/\partial x_a'(l)$ is the canonical momentum.

For the standard model, the Euler-Lagrange equations take the form

$$
R_f \left[\frac{d}{dl} \delta_{ab} + DF_{ab}(\mathbf{x}(l), \boldsymbol{\theta}(l)) \right] \left[\frac{dx_b(l)}{dt} - F_b(\mathbf{x}(l), \boldsymbol{\theta}(l)) \right]
$$

$$
= \frac{[\partial \chi(\mathbf{x}(l) - \mathbf{y}(l)) + 1/2 \nabla_{x(l)} \cdot \mathbf{F}(\mathbf{x}(l), \boldsymbol{\theta}(l))]}{\partial x_a(t)}
$$

$$
\frac{dp_a(l)}{dl} = -DF_{ab}(\mathbf{x}(l), \boldsymbol{\theta}(l)) p_b(l) =
$$

$$
\frac{[\partial \chi(\mathbf{x}(l) - \mathbf{y}(l)) + 1/2 \nabla_{x(l)} \cdot \mathbf{F}(\mathbf{x}(l), \boldsymbol{\theta}(l))]}{\partial x_a(t)}
$$

and we must require

$$
\mathbf{p}_a(l) \frac{\partial F_a(\mathbf{x}(l), \boldsymbol{\theta}(l)}{\partial \boldsymbol{\theta}(l)} = 0, \tag{8.20}
$$

and

$$
DF_{ab}(\mathbf{x}(l), \boldsymbol{\theta}(l)) = \frac{\partial F_a(\mathbf{x}(l), \boldsymbol{\theta}(l))}{\partial x_b(l)}. \tag{8.21}
$$

The E-L equations show how errors represented on the right side of the E-L equation drive the model variables at all layers to move $\mathbf{x}(l) \rightarrow \mathbf{y}(l)$ where data is available.

If one were to specify $\mathbf{x}(l_0)$, but not $\mathbf{x}(l_F)$, then the boundary conditions for the Euler-Lagrange equation are the given $\mathbf{x}(l_0)$ ($\delta \mathbf{x}(l_0) = 0$) and require the canonical momentum $p_a(l_F) = 0$. Examining the Hamiltonian dynamics for this problem then suggests integrating the $\mathbf{x}(l)$ equation forward from l_0 and the canonical momentum equation backward from l_F. This is back propagation.

8.8 Comments on a Set of Curated Retinal Images

We comment on a striking example of the need for investigating the number of samples M required for training a selected network.

In Gulshan et al. (2016) a group of investigators used an existing network that was successful in accurately recognizing image content in another application. This network was trained on 284,335 retinal images from patients diagnosed as healthy or not-healthy as labeled by a group of expert opthamologists. This trained network was "able to predict CV (cardiovascular) risk factors from retinal images with surprisingly high accuracy for patients from two independent data sets of 12,026 and 999 patients" (Peng (2018)).

Were one to have used the tools discussed in this chapter and earlier in Chapter 5, it may have been possible to use many fewer data samples than $M = 284,335$; we do not know. If, however, this were the case, there would be two significant cost and time savings realized:

- this was a supervised learning network. The "supervisors" were trained ophthalmologists who labeled the outputs $\mathbf{y}^k(l_F)$. Fewer supervisors would have meant smaller labor costs.
- The time for training could have been substantially less with a requirement of smaller M.

9

Two Examples of the Practical Use of Data Assimilation

9.1 Data Assimilation in Action

This chapter addresses in some detail in two diverse examples the use of material developed until now.

- It uses the methods to analyze in detail neurobiological data from the laboratory of Daniel Margoliash at the University of Chicago.
- It uses the methods to analyze, in a twin experiment, how many measurements are required to accurately achieve the capability of the fluid dynamical shallow water equations to accurately predict (Pedlosky (1986); Vallis (2017)). The analysis includes Lagrangian drifters.

I have called these analyses 'practical.' In the neurobiological case this allows the user a look at the various 'unobservable' state variables, for example, the voltage dependent gating variables, and an estimation of the many time independent parameters θ. In this sense the DA procedures become a new instrument in the hands of the experimenter enabling the unobservable to be measured through the nonlinear dynamical equations of the biophysics.

If one wishes to build a functional biophysical *network* comprised of the individual birdsong neurons as addressed here, then this provides the ingredients at the nodes of that network. If the model neurons are adopted for the nodes, then use of these models, when validated by their ability to predict, leave the connectivity of the network open for future DA analysis to estimate those synaptic connections among the established neurons.

Once that is accomplished we can address detailed operational questions about the biological network. What we lack at this time is a set of experiments on the network. As we proceed, we may encounter a dearth of computational capacity to carry out the data assimilation tasks, variational or Monte Carlo, we require. Please keep tuned (really no pun intended).

In the matter of the shallow water equations, we describe, via a twin experiment, how one can estimate the number of required measurements one would need to assure accurate predictions following an observation window in time. The shallow water equations are absolutely *not* the full set of dynamical rules required for a global model of the earth system, but the lessons learned in the twin experiment analysis can carry over to the same numerical twin experiments on much larger model to support the design of adequate data collection systems to collect what information is required to complete the model. In these much larger and more detailed models, the shallow water equations are often at the core of the complex dynamics. We address a one-layer set of shallow water equations (An et al. (2017)), those richer models may utilize multiple shallow layers as well as address aspects of the dynamics that go well beyond the fluid flow alone.

The selection of these two rather different application areas emphasizes that the methods presented in this book are universally applicable to many fields, not only to the two examples below. In addition, I trust you will enjoy these examples.

9.2 Experimental Data on Neurons in the Avian Brain

The avian song system is an example of learned functional behavior, namely vocalization, that is passed from generation to generation, overwhelmingly to male birds. The training of juveniles bears a striking resemblance to how humans acquire language. Lessons from the analysis of birdsong have long been thought to illustrate how *natural language processing* is accomplished.

9.2.1 Acquiring the Data

The experiments we rely on are performed *in vitro*. The neurons of interest are in a slice of a region of the avian brain of a songbird. It is called the high vocal center, or HVC, and it contains a few times 10^4 neurons. This slice is immersed in a fluid developed to preserve the activity of the neurons over time. An electrode is inserted into the isolated neuron of interest. It alternates between injecting current into the intracellular medium of the neuron and recording the voltage across the cellular membrane. In the first chapter of the book we showed an example of data from Margolish's lab, see Fig. 0.2, as indicative of the challenges we had before us.

9.2.2 Past Twin Experiments

The results from Twin Experiments, Section 2.6, on individual neurons are rather good and quite encouraging (Toth et al. (2011); Kostuk et al. (2012); Kostuk (2012)). They have provided a test of the data assimilation methods, when asked to produce answers we already know from constructing the data. Here we move on from just testing the DA protocols to using the DA framework to propose and test or validate (or not) models of a neuron's dynamical behavior.

Utilizing laboratory experiments and data assimilation on actual individual neurons to propose and then test biophysical models places more demands on the participants than twin experiments. Among those requirements are these:

- the observed voltages are noisy – that was true in the twin experiments, as we added noise to a model output by hand. In the laboratory, noise on the observed state variable, membrane voltage $V(t)$, may be instrumental or environmental, and it may not be Gaussian. The latter may lead to a reformulation of the 'standard action,' especially in the model error term.
- the stimulus input injected into the neuron is also noisy. This is unavoidable, but with good filters, this noise level may be only 1–2%.
- **we do not know the model**, so we must scour the literature (Senselab-Yale (2020)) to get a good handle on ion currents and ligand gated channels that might have been observed when others have examined the same biophysical neural circuits.
- the neurons into which we inject current, and whose cross membrane voltage response we measure, must be adequately isolated from the rest of the circuit in which they operate in the living animal. This can be achieved. The experimental data we used was acquired *in vitro* where good isolation can be accomplished. One check on this isolation is available when we turn off the stimulating current. We expect to see voltage activity only due to environmental noise and from connections to other neurons. More active voltage response is an indication one is not making the right model for the experiment you are performing, unless you account for those external sources of currents that can affect the neuron you are observing.
- the data was taken in "epochs" of length 2–10 seconds interspersed with rest periods for the neuron. In different epochs, different injected current waveforms were used. This is a useful feature as the intrinsic ion currents (Na, K, Ca,...) are unchanged by the additive $I_{applied}(t)$ in the Hodgkin-Huxley equations (Sterratt et al. (2011); Johnston and Wu (1995)).
- in stark contrast to running twin experiments to test data assimilation methods, in the laboratory neurons wear out and after a few hours of experimentation they die. So the data, strictly speaking, is **not** statistically stationary, which we have actually assumed in the development of the data assimilation methods. This is not a concern if the experimental epochs for presenting specific currents are much shorter than the time constants for changes in the biological preparation associated with multiple injections of current and neuron death.

Current was injected with waveforms selected to meet biophysical criteria we discuss soon, and the voltage was measured at a sampling frequency of 50 kHz – or with a sampling time of $\Delta t = 0.02$ ms. In Fig. 9.1 is an example of the data collected in such an experiment.

From data observed over a time interval $[t_0, t_{final}]$ we wish to estimate the parameters θ and the unobserved voltage gated variables in the class of Hodgkin-Huxley biophysical models we propose for the dynamics of these neurons.

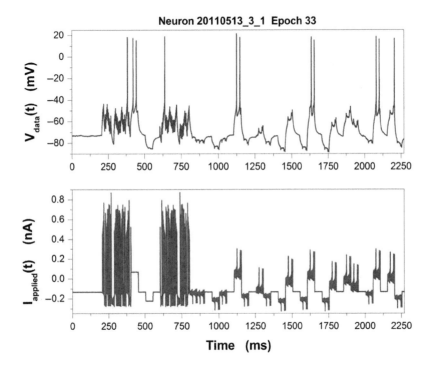

Figure 9.1 **Top Panel** Membrane voltage response of a neuron in the HVC nucleus of a zebra finch to the 2.25 sec of current input shown in the **Bottom Panel**. This region of the avian brain of songbirds is the interface between sensory input through the auditory system and electrical activity controlling the production of song at the songbox. The data are from the laboratory of Daniel Margoliash at the University of Chicago and were acquired along with Daniel Meliza (now at the University of Virginia). The challenge is to use these data, where the only measured state variable is the membrane voltage, to determine the many time independent parameters (of order a few times 10) in a biophysical model of this isolated neuron. One also wants to find the time course for the 10 or so gating variables in the biophysical model that control the flow of various ions into and out of the neuron; these are not observable in experiments, but play a role in the nonlinear dynamics of this neuron's activity. The model adopted for this analysis is discussed in Eqs. (9.1), (9.2), and (9.5).

There are two biophysical criteria placing constraints on the injected current:

- the frequency content of the stimulating current should have little or no power in a range higher than 75–100 Hz. Higher frequencies do not enter the neuron soma, as it has a capacitive time constant that filters out frequencies above that range. Any information in the stimulating current above that range is filtered out.
- the amplitude of the stimulating current should cover a range that drives the neuron across its overall dynamical range so all active channels of current flow in the model of neural activity are stimulated.

These properties of a neuron will be particular to different neurons. So check each one!

A good 'fit' in the estimation window $[t_0, t_{final}]$ does not end the data assimilation task at hand. It may indicate the model is *consistent* with the data. We estimate all parameters and all state variables by the end of the observation window, so we can use the values of $\mathbf{x}(t_{final})$ as initial conditions for integrating the dynamical equations forward for $t > t_{final}$ in a regime where we compare the model output with observations but do not transfer any further information from data to the model. This *prediction window* is the real test of the model we choose to characterize the neuron of interest.

Since we have many epochs of current injection into the same neuron, we may use the model from one epoch to predict the outcome of another epoch. If the two epochs are separated by a long time, the neuron's electrophysiological properties will surely have changed (as noted, the neuron is dying from all this poking about). So we will see below the testing of the neuron's intrinsic characteristics from one epoch to the next.

Sometimes one can use the neuron model from one epoch to predict behavior a few epochs later (or earlier) but, eventually, the nonstationarity of the experimental situation catches up with you.

9.2.3 What Neuron Model Should We Use?

The first thing to try is the NaKL neuron discussed as an example throughout this volume. That might work fairly satisfactorily in temporal regions when the neuron is spiking. It will have to be augmented to allow resolution of observable neuron dynamics. That is usually greater than the few degrees-of-freedom in the simple NaKL model. That's OK, as the competition between inward flowing Na^+ currents causing the membrane potential to rise rapidly and outward flowing K^+ currents bringing the membrane potential back toward its resting potential is the biophysical essence of spiking. However, at voltages well below the threshold for spike formation many other processes are operating. We plan to capture these sub-threshold dynamics as well.

After some consideration we adopt the rather larger Hodgkin-Huxley model involving various Na and K ion currents as well as several Ca^{2+} currents and a curious current called the h-current or HCN-current (Nogaret et al. (2016)).

The first biophysical equation is current conservation:

$$C\frac{dV(t)}{dt} = I_{NaT}(t) + I_{NaP}(t) + I_{K1}(t) + I_{K2}(t) + I_{K3}(t)$$

$$+ I_{CaL}(t) + I_{CaT}(t) + I_{HCN}(t) + I_L(t) + \frac{I_{inj}(t)}{\text{Area}}, \qquad (9.1)$$

The various currents have the Hodgkin-Huxley form (Sterratt et al. (2011); Johnston and Wu (1995)):

$$I_{NaT}(t) = g_{NaT}m_{NaT}(t)^3 h_{NaT}(t)[E_{Na} - V(t)],$$
$$I_{NaP}(t) = g_{NaP}m_{NaP}(t)[E_{Na} - V(t)],$$
$$I_{K1}(t) = g_{K1}n_{K1}(t)^4[E_K - V(t)],$$
$$I_{K2}(t) = g_{K2}m_{K2}(t)^4 h_{K2}(t)[E_K - V(t)],$$
$$I_{K3}(t) = g_{K3}m_{K3}(t)[E_K - V(t)]$$
$$I_{CaL}(t) = g_{CaL}m_{CaL}(t)^2 I_{GHK}(t),$$
$$I_{CaT}(t) = g_{CaT}m_{CaT}(t)^2 h_{CaT}(t)I_{GHK}(t),$$
$$I_{HCN}(t) = g_{HCN}h_{HCN}(t)[E_{HCN} - V(t)],$$
$$I_L(t) = g_L[E_L - V(t)]. \qquad (9.2)$$

The maximal conductances of the various ion channels $\{g_{NaT}, g_{NaP}, g_{K1}, g_{K2}, g_{K3}, g_{CaL}, g_{CaT}, g_{HCN}, g_L\}$ set the scale of current flow into or out of the neuron. We use the *nominal* values in units of $\frac{mS}{cm^2}$: $\{g_{NaT} = 110, g_{NaP} = 0.064, g_{K1} = 5, g_{K2} = 15, g_{K3} = 9.1, g_{HCN} = 0.092, g_L = 0.06\}$. C is the membrane capacitance per unit area, $I_{current}(t)$ are the current densities of the ion channels. 'Area' is the surface area of the neuron membrane through which the external current is injected.

The Ca^{2+} currents use a voltage dependence adapted to the fact that the intracellular concentration of $[Ca^{2+}]_{in}$ is approximately 10^{-4} of the extracellular $[Ca^{2+}]_{out}$ concentration; this called the Goldman-Hodgkin-Katz (GHK) current (Sterratt et al. (2011)):

$$I_{GHK}(t) = \frac{V(t)}{(1 - \exp[-V(t)/V_T])}\left[[Ca^{2+}]_{in} - [Ca^{2+}]_{out}\exp[-V(t)/V_T]\right],$$
$$(9.3)$$

in which $V_T = RT/2F \approx 12.5$ mV at room temperature. F is the Faraday constant, 96,485 C/mol, R is the gas constant, 8.3 J/(mol K), and T is the temperature in degrees Kelvin at the cell. At room temperature $RT/F \approx 25.7$mV.

The reversal potentials (Sterratt et al. (2011); Johnston and Wu (1995)), or the Nernst potentials, are approximately (in mV): $\{E_{Na} = 40 - 55, E_K = (-90) - (-80), E_L = (-50) - (-60), E_h = E_{HCN} = -37\}$.

Each of the gating variables

$$a(t) = \{m_{NaT}(t), h_{NaT}(t), m_{NaP}(t), n_{K1}(t), n_{K2}(t),$$

$$n_{K3}(t), m_{CaL}(t), m_{CaT}(t), h_{CaT}(t), h_{HCN}(t)\} \tag{9.4}$$

is taken to satisfy a first order kinetics equation of the form

$$\frac{da(t)}{dt} = \frac{a_0(V(t)) - a(t)}{\tau_a(V(t))}, \tag{9.5}$$

which is discussed in Sterratt et al. (2011); Nogaret et al. (2016). $0 \leq a(t) \leq 1$.

This biophysical Hodgkin-Huxley model has $D = 11$ state variables and $N_p \approx 70$ time independent parameters. We have $L = 1$ observed state variable. In the data assimilation procedure one uses the experimental knowledge of $I_{inj}(t)$, which is selectable by the observer, and $V(t)$, which is observed by the observer, along with one of the data assimilation methods discussed in this volume. In the results reported here, we used "old nudging."

From numerical simulations via twin experiments on simpler neuron models, we anticipate that if a current in the model has maximal conductance g_{model}, but that current does *not* appear in the data, then $g_{model} \approx 0$. So the data will prune the model.

This pruning appeared in a set of twin experiments where data from a neuron without an h-current was presented to a model with an h-current present. The other currents had maximal conductances of order 1, and the h-current had $g_h \approx 2 \times 10^{-9}$ (Toth et al. (2011)).

The data assimilation procedures we know about as of early 2020 do not (yet) suggest currents to **add** to the model, if needed. That is still part of the art of modeling neural systems.

The use of neuron parameters θ estimated in one epoch (here, Epoch 33) to predict in the next epoch (here, 34) shows the consistency of the model used in Epoch 34 with its development in Epoch 33.

9.2.4 Implications of the Successful Use of SDA in Data from Laboratory Experiments

This gallery of results emphasizes the three ingredients we need for data assimilation and assessing the validity of the physical model chosen for the nonlinear dynamics of the neuron in any epoch of observation (used for estimation when allied to a data assimilation method) and epochs of prediction using models selected by observations and data assimilation in nearby epochs:

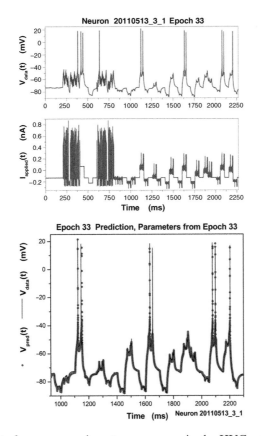

Figure 9.2 Data from an experiment on a neuron in the HVC nucleus of a song
bird – a zebra finch in the Margoliash laboratory at the University of Chicago.
These data come from Epoch 33 in the experiment. In the **Upper Sub-panel** we
display the voltage response to the injected current (**Lower Sub-panel**) over a
time period of 2.25 sec. In the **Lower Panel** we show the prediction of the biophysical Hodgkin-Huxley model constructed for the dynamics of this neuron. In
a 900ms observation window the time independent parameters θ and the time
dependent *unobserved* gating variables of the model were estimated. The values
at $t_{final} = 900$ ms were used to integrate the Hodgkin-Huxley model forward
to 2225 ms. The observed voltage data for $t \geq t_{final}$ are shown in black, and
the model predictions are shown in blue. The model produces an extra spike near
1650 ms.

- *Data*, which we have acquired with knowledge of the instrument injecting
  current (the forcing of the system) into the neuron.
- *A Model*, which we have selected by scouring the Senselab Neuron Database (Senselab-Yale (2020)) for an indication of currents appropriate to the
  neuron we are investigating, and
- *A tested data assimilation method*, which we tested by using it on models with
  the same architecture as we propose to use on laboratory data we collect.

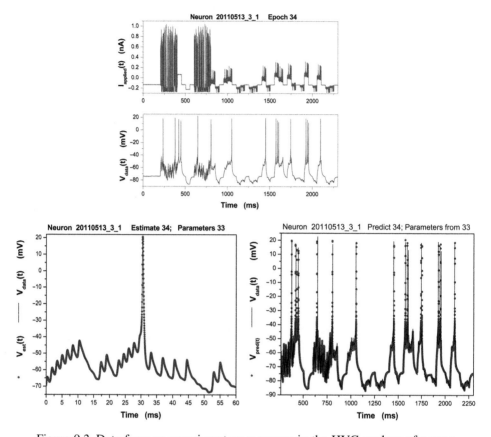

Figure 9.3 Data from an experiment on a neuron in the HVC nucleus of a song bird: a zebra finch in the Margoliash laboratory at the University of Chicago. These data come from Epoch 34 in the experiment. In the **Bottom Sub-panel** we display the voltage response to the injected current (**Top Sub-panel** over a time period of 2.26 sec). The voltage response data was sampled at 50 kHz for 900 ms of the 2250 ms time course. In the **Lower Right Panel** we show the prediction of the voltage time course in Epoch 34 over $0\,\text{ms} \leq t \leq 60\,\text{ms}$ using the biophysical model constructed in Epoch 33, but the injected current used in Epoch 34. In the **Lower Left Panel** we display the prediction of the response to the injected current in Epoch 34 $250\,\text{ms} \leq t \leq 2260\,\text{ms}$ using the biophysical model constructed in Epoch 33. The injected currents in Epochs 33 and 34 are different in detail. The neuron is the same, but one epoch older in Epoch 34.

In each of the examples we estimate all the unknown time independent parameters θ and all of the unobserved state variables throughout the observation window $[t_0, t_{final}]$. But this is not the end of our work on each data set. We have only one way to test the quality of the model we proposed. We follow that 'fit' to the data and use the estimated parameters and the unobserved state variables at t_{final} as initial conditions to integrate the proposed dynamical equations forward in time:

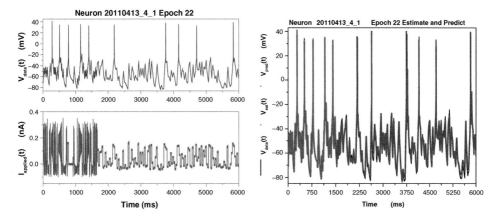

Figure 9.4 Data from an experiment on a neuron in the HVC nucleus of a song bird: a zebra finch in the Margoliash laboratory at the University of Chicago. These data come from Epoch 22 in the experiment. In the **Left Top Sub-panel** we display the voltage response to the injected current **Left Bottom Sub-panel** over a time period of 6.0 sec. The voltage response data was sampled at 50 kHz for 1600 ms of the 6000 ms time course. In the **Right Panel** we show the estimation of the voltage time course in Epoch 22 over $0 \, \text{ms} \leq t \leq 1600 \, \text{ms}$ informing the biophysical model constructed in Epoch 22. We also display the prediction of the response to the injected current in Epoch 22 $1600 \, \text{ms} \leq t \leq 6000 \, \text{ms}$ using the biophysical model constructed in Epoch 22.

$t > t_{final}$ to test the prediction (called generalization in machine learning contexts); this examines the validity of the model proposed to understand the data. It is a test of the quality of the estimation of the θ and of the unobserved state variables.

In the analysis of the experiments shown in Fig. 9.2–9.9, we predict often in the same epoch in which the data was acquired, and we predict in a nearby epoch using the model (parameters θ) constructed in another nearby epoch. In Fig. 9.9, for example, we predict the neuron's voltage response to the current waveform presented in Epoch 12 to the model built using the Epoch 11 data. The accuracy and consistency of the times when we predict within and without the same epoch is a demanding test of the model proposed, and, actually, of the data assimilation method as well.

9.3 Shallow Water Equations; Lagrangian Drifters

The problem of forecasting the behavior of a complex dynamical system through analysis of observational time-series data becomes difficult when the system expresses chaotic behavior and the measurements are sparse, in both space and/or time. Despite the fact that this situation is quite typical across many fields, including numerical weather prediction, the issue of whether the available observations

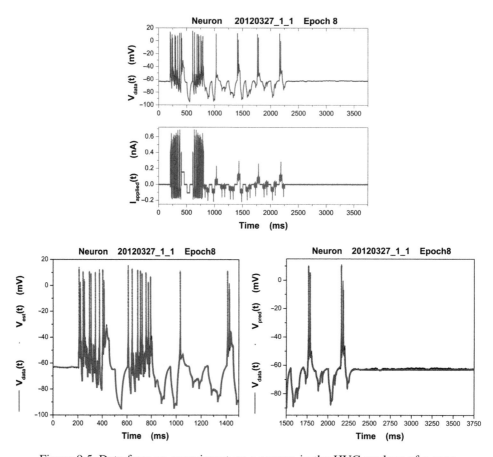

Figure 9.5 Data from an experiment on a neuron in the HVC nucleus of a song bird: a zebra finch in the Margoliash laboratory at the University of Chicago. These data come from Epoch 8 in the experiment. In the **Top Sub-panel** we display the voltage response to the injected current (**Bottom Sub-panel** over a time period of 3.75 sec). The voltage response data was sampled at 50 kHz for 1500 ms of the 3750 ms time course. In the **Right Panel** we show the estimation of the voltage time course in Epoch 8 over $0\,\text{ms} \leq t \leq 1500\,\text{ms}$ informing the biophysical model constructed in Epoch 8. In the **Left Panel** we display the prediction of the response to the injected current in Epoch 8 $1500\,\text{ms} \leq t \leq 3750\,\text{ms}$ using the biophysical model constructed in Epoch 8.

are 'sufficient' for generating successful forecasts is still not well understood or even addressed in most work using the available sensors (An et al. (2017)).

An analysis by Whartenby et al. (2013) found that in the context of the non-linear shallow water equations on a β-plane, standard nudging techniques require observing approximately 70% of the full set of state variables.

We look now at the same system using a method introduced by Rey et al. (2014a); Rey (2017), which generalizes standard nudging methods to utilize

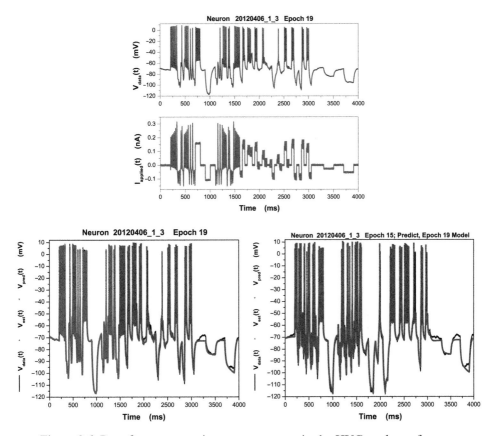

Figure 9.6 Data from an experiment on a neuron in the HVC nucleus of a song bird: a zebra finch in the Margoliash laboratory at the University of Chicago. These data come from Epoch 19 in the experiment. In the **Top Panels** we display the voltage response to the injected current over a time period of 4.0 sec. The voltage response data was sampled at 50 kHz for 1600 ms of the 4000 ms time course. In the **Lower Panels** we show the estimation of the voltage time course in Epoch 19 over $0\,\text{ms} \leq t \leq 1600\,\text{ms}$ informing the biophysical model constructed in Epoch 19. **Lower Left Panel** We display the prediction of the response to the injected current in *Epoch 15* $1600\,\text{ms} \leq t \leq 4000\,\text{ms}$ using the biophysical model constructed in Epoch 19. Note that we do not know the initial conditions when we start Epoch 15, but we do know the parameters. Look closely at the small red area in the **Lower Right Panel** where we exercise our data assimilation capability over 100 ms to find a good estimate of the initial conditions for predicting in Epoch 15.

time delayed measurements. This was discussed earlier in Chapter 4, and we liberally use results from there. We show that in certain circumstances, waveform information provides a sizable reduction in the number of observations required to construct accurate estimates and high-quality predictions. In particular, we find that this estimate of 70% can be reduced to about 33% using

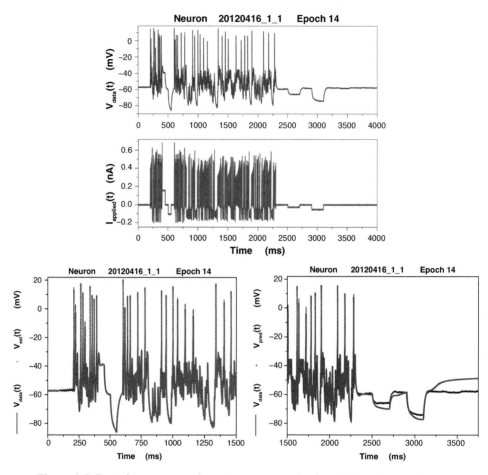

Figure 9.7 Data from an experiment on a neuron in the HVC nucleus of a song bird: a zebra finch in the Margoliash laboratory at the University of Chicago. These data come from Epoch 14 in the experiment. In the **Top Panels** we display the voltage response to the injected current over a time period of 4.0 sec. The voltage response data was sampled at 50 kHz for 1500 ms of the 4000 ms time course. In the **Lower Left Panel** we show the estimation of the voltage time course in Epoch 14 over 0 ms $\leq t \leq$ 1500 ms informing the biophysical model constructed in Epoch 19. **Lower Right Panel** We display the prediction of the response to the injected current in Epoch 14 1500 ms $\leq t \leq$ 4000 ms using the biophysical model constructed in Epoch 14.

time delays, and even further if Lagrangian drifter locations are also used as measurements.

Operational Numerical Weather Prediction (NWP) models at the European Centre for Medium-Range Weather Forecasting (ECMWF) now contain upwards of 10^{10} degrees of freedom once the partial differential equations of fluid dynamics have been put on a spatial grid representing the atmosphere and in the ocean, and

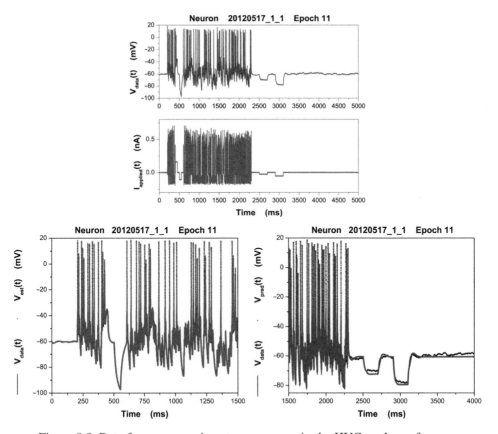

Figure 9.8 Data from an experiment on a neuron in the HVC nucleus of a song bird: a zebra finch in the Margoliash laboratory at the University of Chicago. These data come from Epoch 11 in the experiment. In the **Top Panels** we display the voltage response to the injected current over a time period of 4.0 sec. The voltage response data was sampled at 50 kHz for 1500 ms of the 4000 ms time course. In the **Lower Left Panel** we show the estimation of the voltage time course in Epoch 11 over $0 \, \text{ms} \leq t \leq 1500 \, \text{ms}$ informing the biophysical model constructed in Epoch 19. **Lower Right Panel** We display the prediction of the response to the injected current in Epoch 11 $1500 \, \text{ms} \leq t \leq 4000 \, \text{ms}$ using the biophysical model constructed in Epoch 11.

the Physics of subgrid scale phenomena have been introduced as well. These models are analyzed through methods of data assimilation using $3 - 5 \times 10^8$ daily observations. Given the scale of these calculations, the question of whether the remaining observations are in fact 'sufficient' for producing accurate analyses and forecasts is of considerable practical importance.

The twin experiment analysis by Whartenby et al. (2013) showed that standard "nudging-old" methods, when applied to a chaotic, shallow water flow on a β-plane driven by Ekman pumping (Pedlosky (1986); Vallis (2017)),

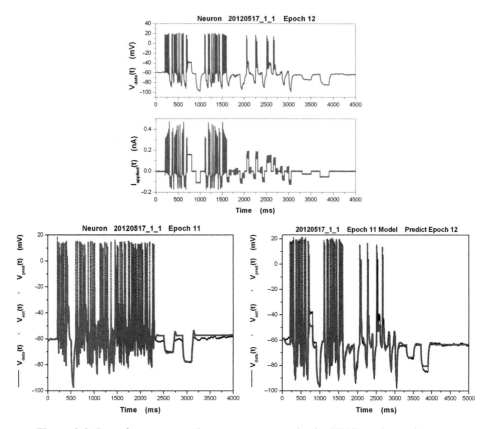

Figure 9.9 Data from an experiment on a neuron in the HVC nucleus of a song bird: a zebra finch in the Margoliash laboratory at the University of Chicago. These data come from Epoch 12 in the experiment. In the **Top Panels** we display the voltage response to the injected current over a time period of 4.5 sec. The voltage response data was sampled at 50 kHz for 1500 ms of the 4500 ms time course. In the **Lower Left Panel** we show the estimation of the voltage time course in Epoch 11 over $0\,\mathrm{ms} \leq t \leq 1500\,\mathrm{ms}$ informing the biophysical model constructed in Epoch 11. **Lower Right Panel** We display the prediction of the response to the injected current in Epoch 12 $1500\,\mathrm{ms} \leq t \leq 4000\,\mathrm{ms}$ using the biophysical model constructed in Epoch 11. Again, notice the 100 ms time window used to estimate the initial state of the neuron driven by the Epoch 12 current when the neuron model is from Epoch 11.

required direct observation of roughly 70% of the $3N_\Delta^2$ dynamical variables, $\{u(\mathbf{r}, t), v(\mathbf{r}, t), h(\mathbf{r}, t)\}\,\mathbf{r} = \{x, y, z\}$, on a spatial grid of size $N_\Delta \times N_\Delta$ to achieve accurate forecasts. Predictions were poor unless the height variable $h(\mathbf{r}, t)$ and at least one of the two velocity variables $u(\mathbf{r}, t)$, $v(\mathbf{r}, t)$ at each of the N_Δ^2 grid points were measured.

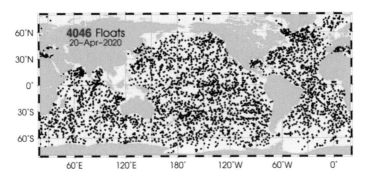

Figure 9.10 Lagrangian drifters world wide as of April 20, 2020

This threshold was termed the critical number of measurements L_c required to synchronize the model with the data. It depends on a number of factors, including the type of observation network, the signal to noise ratio in the observations, properties of the model such as the number and magnitude of its Lyapunov exponents, as well as the choice of integration and DA algorithm. Strong constraint 4DVar for instance, which is now standard practice in data assimilation (Rabier et al. (1994)), encounters serious difficulty when the length of the window is long relative to the timescale of the chaos (Pires et al. (1996)). In this case, the algorithm will not produce adequate forecasts even with full observations $L = D$. Despite this, however, the lowest estimates of L_c appear remarkably consistent using different nudging methods and fixed interval formulations of 4DVar with both strong and weak constraints (Abarbanel et al. (2009); Abarbanel (2013); Quinn (2010)).

What can be done when $L < L_c$?

Specifically, we will show that using the method introduced by Rey et al. (2014a,b), which modifies a standard nudging technique to include additional information in the *time delays* of the observations, namely using the information in the waveforms of the data, the estimate of 70% given by Whartenby et al. (2013) can be reduced to roughly 33%. These outcomes suggest that time delays may be useful for reducing the number of required observations to meet the practical constraints of operational Numerical Weather Prediction.

9.4 Time-Delayed Nudging Used in the Shallow Water Equations

This is covered in some detail in Chapter 4; however, we find repetition may have pedagogical merit.[1] We now briefly describe the concept of time delayed nudging. Further details can be found in Rey et al. (2014a,b). The system of interest is

---

[1] "Punitive Pedagogy" is the practice of presenting all ideas once, and only once, no matter how important they may be. We try not to employ this method.

assumed to be described by a mathematical model, whose state is given by a D-dimensional vector $\mathbf{x}(t)$. The model defines a dynamical rule for evolving the $\mathbf{x}(t)$ in time, which we assume can be represented as a set of ordinary differential equations (ODEs):

$$\frac{d\mathbf{x}(t)}{dt} = \mathbf{F}(\mathbf{x}(t), t). \tag{9.6}$$

If the dynamics of the system are described by partial differential equations (PDEs), such as with fluids in an earth systems model, these ODEs may be realized by discretizing the PDEs on a spatial grid or the equivalent. It is possible that discretization error may be introduced in this process.

The measurements $\mathbf{y}(\tau_k)$ are made at a subset L of the $D = 3N_\Delta^2$-dimensional state variables $\mathbf{x}(t)$. $y_l(t) = x_l(t) + \text{noise } l = 1, 2, \ldots L$.

We wish to estimate the full model state $\mathbf{x}(t_{final})$ at the end of the assimilation window and all fixed time parameters $\boldsymbol{\theta}$ using information from observations. Then we intend to use these estimates to predict the system's subsequent behavior for $t > t_{final}$ using the model dynamics (Eq. (9.6)). The accuracy of these predictions, when compared with additional measured data in the prediction window $t > t_{final}$, where no additional information from observations is received by the model, serves as a metric to validate both the model and the data assimilation method, through which the unobserved states of the system are determined. This establishes a necessary condition on L that is required to synchronize the model output with the data and thereby obtain accurate estimates for the unobserved states of the system.

When the model is known precisely, a familiar strategy for transferring information from the measurements to the model involves the addition of a coupling or control or nudging term to Eq. (9.6),

$$\frac{d\mathbf{x}(t)}{dt} = \mathbf{F}(\mathbf{x}(t), t) + \mathbf{G}(t) \cdot \big(\mathbf{y}(t) - \mathbf{x}(t)\big). \tag{9.7}$$

$\mathbf{G}(t)$ is an $D \times L$ matrix that is nonzero only at times $t = \tau_k$ where measurements occur. This long-standing procedure, known as 'nudging' in the geophysics and meteorology literature, has been shown to fail when the number of measurements at a given observation time is smaller than a critical value L_c (Abarbanel et al. (2009)). This can be understood by noting that the coupling term perturbs the observed model states, driving them toward the data. With enough observations L, and a sufficiently strong coupling $\mathbf{G}(t)$, this control term alters the Jacobian of the dynamical system (Eq. (9.7)) so that all its conditional Lyapunov exponents are negative (Pecora and Carroll (1990); Abarbanel (1996); Kantz and Schreiber (2004); Kostuk (2012)). That is, the log of the maximum eigenvalue of the matrix $[\Phi(t_{final}, t_0)^T \cdot \Phi(t_{final}, t_0)]^{1/2T}$ is negative, where $\Phi(t, t')$ is the solution to the variational equation

$$\frac{d\Phi(t, t')}{dt} = D\mathbf{F}(\mathbf{x}(t), t) \cdot \Phi(t, t') \qquad \Phi_{ab}(t, t) = \delta_{ab}, \qquad (9.8)$$

along the trajectory given by Eq. (9.7) and $D\mathbf{F} = DF(\mathbf{x}(t), t) - \mathbf{G}(t)$ is its Jacobian.

It is, therefore, important to understand, for a given problem, whether $L > L_c$. If this condition is not satisfied and additional measurements cannot be made, then we must find another means to overcome this deficit in L.

One way to proceed involves the recognition that additional information resides in the temporal derivatives of the observations. In practice, however, this derivative information cannot be measured directly, although it can be approximated via finite differences, for instance by approximating $d\mathbf{y}(t_n)/dt$ with $(\mathbf{y}(t_n + \tau) - \mathbf{y}(t_n))/\tau$ where τ is some multiple of the time differences between measurements. The drawback here is that the derivative operation acts as a high-pass filter, and is thus quite susceptible to noise in the measurements.

Alternatively, it has been known for some time in the nonlinear dynamics literature that this additional information in the derivative is also available in the *time delays* (Abarbanel (1996); Kantz and Schreiber (2004)) of the measurements, $\mathbf{y}(t_n + \tau)$. This process can be repeated as many times as needed to form a D_M dimensional vector of time delays, which we call $\mathbf{S}(t)$.

This idea provides the basis for the well-established technique in the analysis of nonlinear dynamical systems, where it is employed as a means of reconstructing unambiguous orbits of a partially observable system (Aeyels (1981a,b); Mañé (1981); Sauer et al. (1991); Takens (1981); Kantz and Schreiber (2004); Abarbanel (1996)). By mapping to a proxy space of time delayed observations, one is able to invert the projection associated with measuring $L < D$ components of the underlying dynamics, by using the fact that new information beyond $\mathbf{y}(t_n)$ is to be found in $\mathbf{y}(t_n + \tau)$.

Note that the *time delay* τ and the *embedding dimension* D_M are parameters that need to be chosen appropriately for the system, although a number of useful heuristics are available (Abarbanel (1996); Kantz and Schreiber (2004)). Takens (1981) proved that that taking $D_M > 2 D_A$, where D_A is the fractal dimension of the attractor, is sufficient to unambiguously reconstruct the topology of the attractor. It is worth noting, however, that this condition is only sufficient, and the procedure often succeeds with considerably a smaller value of D_M. In the estimation context, the time delays are used in a slightly different way. Instead of reconstructing the topology of the attractor, they are used in this context to control local instabilities in the dynamics. In other words, D_M does not need to embed the entire space. Rather, it only needs to be large enough to effectively increase the amount of information transferred from the L measurements to a value above the critical threshold, L_c.

Using this idea Rey et al. (2014a,b) proposed a technique to extract additional information from time delayed observations by constructing an extended state space $\mathbf{S}(t)$, created from an $L \cdot D_M$ dimensional vector of the measurements and its time delays. The components of this time delayed observation vector are denoted by

$$\mathbf{Y}(t_n) = \{\mathbf{y}(t_n), \mathbf{y}(t_n + \tau), \ldots, \mathbf{y}(t_n + \tau (D_M - 1))\}, \qquad (9.9)$$

where D_M is the dimension of the time delayed vector $\mathbf{Y}(t_n)$, and τ is the delay, which here is assumed to be an integer multiple of Δt. The corresponding time delay model vectors $\mathbf{S}(\mathbf{x}(t))$ are given by

$$\mathbf{S}(\mathbf{x}(t)) = \{[\mathbf{x}(t)], [\mathbf{x}(t + \tau)], \ldots, [\mathbf{x}(t + \tau (D_M - 1))]\}, \qquad (9.10)$$

where the values $\mathbf{x}(t' > t)$ are constructed by integrating the *uncoupled* dynamics, Eq. (9.6), forward in time. The time evolution for $\mathbf{S}(\mathbf{x}(t))$ is given by the chain rule,

$$\frac{d\mathbf{S}(\mathbf{x}(t))}{dt} = \mathbf{DS}(\mathbf{x}(t)) \cdot \mathbf{F}(\mathbf{x}(t), t), \qquad (9.11)$$

where the Jacobian $\mathbf{DS}(\mathbf{x}(t)) = \partial \mathbf{S}(\mathbf{x}(t))/\partial \mathbf{x}(t)$ with respect to $\mathbf{x}(t)$ can be computed using the variational Eq. (9.8), by substituting the Jacobian of the uncoupled model $\{DF\} \to \mathbf{DF}$. Furthermore, in analogy with Eq. (9.7), we introduce a control term $\mathbf{g}(t)$ in time delay space. We choose $\mathbf{g}(t)$ to be the unit matrix with no loss of generality.

$$\frac{d\mathbf{S}(\mathbf{x}(t))}{dt} = \mathbf{DS}(\mathbf{x}(t)) \cdot \mathbf{F}(\mathbf{x}(t), t) + \big(\mathbf{Y}(t) - \mathbf{S}(\mathbf{x}(t))\big). \qquad (9.12)$$

We then transform back to physical space, by multiplying both sides of this equation by $[\mathbf{DS}(\mathbf{x}(t))]^{-1}$, to arrive back in physical space:

$$\frac{d\mathbf{x}(t)}{dt} = \mathbf{F}(\mathbf{x}(t), t) + \mathbf{G}(t) \cdot [\mathbf{DS}(\mathbf{x}(t))]^{-1}\big(\mathbf{Y}(t) - \mathbf{S}(\mathbf{x}(t))\big). \qquad (9.13)$$

Since $\mathbf{DS}(\mathbf{x}(t))$ is an $(L \cdot D_M) \times D$ matrix, it is generally not square so its pseudoinverse $[\mathbf{DS}(\mathbf{x}(t))]^+$ is used.

At each step of the integration of the controlled (nudged) dynamical equations (Eq. (9.13)), the control term perturbs the full state vector in time delay space $\mathbf{S}(\mathbf{x}(t))$ toward the time delay measurement vector $\mathbf{Y}(t)$, allowing it to extract additional information from the waveform of the *existing measurements*. The value of this statement will become clear as we proceed.

In the limit $D_M = 1$ the time delay formulation Eq. (9.13) reduces to the standard nudging control Eq. (9.7). Two important differences, however, are realized when $D_M > 1$.

- First, information from the time delays of the observations is presented to the physical model equations, and
- second, all components of the model state $\mathbf{x}(t)$ are influenced by the control term, not just the observed components. This, for example, allows fixed parameters $\boldsymbol{\theta}$ of the model to be estimated as a natural result of the synchronization process by including them as additional state variables, satisfying $d\boldsymbol{\theta}(t)/dt = 0$.

It is also worth pointing out that here we are not using time 'delays' in their usual sense, but rather a *time advanced* formulation that looks forward in time. The motivation for this is related the idea of assimilation in the unstable subspace (Trevisan et al. (2010); Palatella et al. (2013)), where the goal is to control the propagation of errors on the unstable manifold. Since these errors are locally described by Eq. (9.8) as the system evolves forward in time, the time advanced construction is a natural choice. Both formulations are acceptable, however. This also brings up a concern regarding what to do at the end of the assimilation window. One option is to switch to a time delayed formulation, or perhaps a mixed formulation that uses delays both forwards and backwards in time. These issues will not be considered here, however. Instead, our numerical experiments use only a time advanced formulation, by choosing the end of the observation window so that the last observation $\mathbf{y}(t_{final} + \tau (D_M - 1))$ is always available.

Time delay nudging shares considerable overlap with incremental formulations of strong constraint 4DVar (Lewis and Derber (1985); Talagrand and Courtier (1987)). For instance, both methods use a sliding window of observations and compute the control (nudging) perturbation by minimizing the magnitude of the time-distributed innovations $|\mathbf{Y}(t) - \mathbf{S}(\mathbf{x}(t))|^2$. The main differences are these: (i) 4DVar does not include the notion of a time delay or embedding dimension; (ii) with the time delay method, the observation window moves in small increments Δt, so observations are reused in multiple analyses, and (iii) time delay nudging using truncated singular-value decomposition, while strong constraint 4DVar uses a background term to regularize the solution.

9.5 Twin Experiments

We proceed to test our time delay nudging procedure through a series of *twin experiments* (Blum et al. (2009); Blum (2010)). After solving the original dynamical equations Eq. (9.6) forward from a selected initial condition $\mathbf{x}(0)$, the observed data is taken as the projection down to the L observed components. Gaussian noise $N(0, \sigma)$ is added to each component to simulate observation error.

We monitor our progress as time goes by via calculating the observable synchronization error, namely the root mean square deviation between the data and the observed model states,

$$SE(t_n) = \sqrt{\frac{1}{L} |\mathbf{x}^s(t_n) - \mathbf{y}^s(t_n)|^2}, \tag{9.14}$$

where scaled variables have been introduced such that $x_\ell^s(t) = [x_\ell(t) - x_\ell^{min}(t)]/[x_\ell^{max}(t) - x_\ell^{min}(t)]$ and $x_\ell^{min/max}(t)$ are the minimum or maximum values of $x_\ell(t)$ over the entire assimilation window. The same definition holds for $y_\ell^s(t)$. This rescales all data and observed model states to lie in the interval [0, 1], so that each state component's contribution to the synchronization error is roughly equal. While this could make the result sensitive to outliers in the data, it does not appear to be an issue.

When the estimation is completed at time $t = t_{final}$ the coupling term $\mathbf{G}(t)$ goes to zero, and the uncoupled dynamics (Eq. (9.6)) are integrated forward from the estimated $\mathbf{x}(t_{final})$ and estimated fixed time parameters $\boldsymbol{\theta}$ to construct a forecast for $t > t_{final}$. This may be compared with additional observations $\mathbf{y}(t > t_{final})$.

Comparing the forecast to observations made when $t \geq t_{final}$ provides confidence that the unobserved state variables are also accurately estimated at t_{final}.

It was previously shown by Whartenby et al. (2013) that when the synchronization error (Eq. (9.14)) decreases to very small values, the full state $\mathbf{x}(t_{final})$ is accurately estimated and the forecast is quite good. Conversely, when this fails to occur, the full state $\mathbf{x}(t_{final})$ is not well estimated and the prediction is unreliable.

In Rey et al. (2014a,b), this contraction of the synchronization error was only observed when the number of time delayed observations $L \times D_M$, and the magnitude of the coupling matrix $\mathbf{G}(t)$ was 'large enough.' The precise meaning of this statement will become apparent as we proceed.

9.6 Nonlinear Shallow Water Equations

We now describe the application of time delay nudging to a nonlinear model of shallow water flow on a mid-latitude β-plane. This geophysical fluid dynamical model (previously examined by Pedlosky (1986); Vallis (2017); and Whartenby et al. (2013), among many others) is at the core of earth system flows used in Numerical Weather Prediction. Of course, operational models contain considerably more geophysical detail than our example here. Those models also describe the dynamics over a sphere. While we suspect the results presented here for this simplified model will be applicable to more realistic models as well, additional numerical experiments are necessary to validate this claim.

As the depth of the coupled atmosphere ocean fluid layer (10–15 km) is markedly less than the earth's radius (6400 km), the shallow water equations for two-dimensional flow as derived in Pedlosky (1986) provide a good approximation to the fluid dynamics of the ocean and atmosphere. Three fields on a mid-latitude plane describe the fluid flow $\{u(\mathbf{r}, t), v(\mathbf{r}, t), h(\mathbf{r}, t)\}$: the north-south velocity $v(\mathbf{r}, t)$, the east-west velocity $u(\mathbf{r}, t)$, and the height of the fluid $h(\mathbf{r}, t)$, with $\mathbf{r} = \{x, y\}$. The fluid is taken as a single, constant density layer and is driven by wind stress $\tau_s(\mathbf{r}, t)$ at the surface $z = h(\mathbf{r}, t)$ through an Ekman layer. These physical processes satisfy the following dynamical equations with $\mathbf{u}(\mathbf{r}, t) = \{u(\mathbf{r}, t), v(\mathbf{r}, t)\}$ (Pedlosky (1986)):

$$\frac{\partial \mathbf{u}(\mathbf{r}, t)}{\partial t} = -\mathbf{u}(\mathbf{r}, t) \cdot \nabla \mathbf{u}(\mathbf{r}, t) - g \nabla h(\mathbf{r}, t) + \mathbf{u}(r, t) \times f(y)\hat{z}$$
$$+ A \nabla^2 \mathbf{u}(\mathbf{r}, t) - \epsilon \, \mathbf{u}(\mathbf{r}, t)$$
$$\frac{\partial h(\mathbf{r}, t)}{\partial t} = -\nabla \cdot \left[h(\mathbf{r}, t) \, u(\mathbf{r}, t) \right] - \hat{z} \cdot \text{curl} \left[\frac{\tau_s(\mathbf{r}, t)}{f(y)} \right]. \tag{9.15}$$

The Coriolis force is linearized about the equator $f(y) = f_0 + \beta y$ and the wind-stress profile is selected here to be $\tau_s(\mathbf{r}, t) = [F/\rho] \cos(2\pi \, y), 0\}$. The parameter A represents the viscosity in the shallow water layer, ϵ is Rayleigh friction, and $\hat{z}$ is the unit vector in the z-direction. The values we have used for the model parameters are given in Table 1. With these fixed parameters the shallow water flow is chaotic, and the largest Lyapunov exponent (Abarbanel (1996); Kantz and Schreiber (2004)) for this flow is estimated to be $\lambda_{max} = 0.0325/\text{hour} \approx 1/31 \, \text{hour}$ by measuring the average growth rate of random perturbations.

The shallow water flow was determined using the entropy conserving discretization scheme given by Sadourny (1975) on a grid of size $N_\Delta \times N_\Delta$ for increasing resolution $N_\Delta = \{16, 32, 64\}$ over periodic boundary conditions. Using the twin-experiment framework, with simple nudging given in Eq. (9.7) and a static, uniform observation operator, approximately 70% of the $D = 3N_\Delta^2$ degrees of freedom were required to synchronize the model output with the data (Whartenby et al. (2013)). In other words, the height field and at least one of the velocity fields at each grid point was required to be observed.

Since the results we found were roughly consistent using the three distinct resolutions we used, $N_\Delta = 16, 32, 64$, we restrict our discussion here to the case where $N_\Delta = 16$. So the total number of degrees of freedom is $D = 3N_\Delta^2 = 768$, for which Whartenby et al. (2013) estimated $L_c \approx 524 = 0.68 \, D$.

9.7 Results with Time Delay Nudging for the Shallow Water Equations

We now demonstrate that the time delay method is capable of reducing L_c, by showing that it can construct successful estimates and predictions without directly observing the horizontal velocity fields. This strategy was shown to fail by Whartenby et al. (2013) with static ($D_M = 1$) nudging. Thus, we assume height

Table 9.1 *Parameters used in the generation of the shallow water 'data' for the twin experiment. All fields as well as {x, y, t} were scaled by the values in the table, so all calculations were done with dimensionless variables.*

Parameter	Physical Quantity	Value in Twin Experiments
Δt	Time Step	36 s
ΔX	East-West Grid Spacing	50 km
ΔY	North-South Grid Spacing	50 km
H_0	Equilibrium Depth	5.1 km
f_0	Central value of the Coriolis parameter	5×10^{-5} s^{-1}
β	Meridional derivative of the Coriolis parameter	2×10^{-11} m^{-1}s^{-1}
F/ρ	Wind Stress	0.2 m^{2}s^{-3}
A	Effective Viscosity	10^{-4} m^{2}s^{-1}
ϵ	Rayleigh Friction	2×10^{-8} s^{-1}

measurements alone are made at each grid point (i, j) for $i, j = \{1, 2, \ldots, 16 = N_\Delta\}$, so $L = 256 < 524 \approx L_c$, as estimated by Whartenby et al. (2013).

The coupling matrix $\mathbf{G}(t)$ is taken to be diagonal, with different weights for the heights and for the velocities. In particular, $G_{u,v} \Delta t = 0.5$ and $G_h \Delta t = 1.5$ with $\Delta t = 0.01$ hour $= 36$ s. The values of G_h are larger than G_u, G_v, since the average height is 5000 ± 30 m, three orders of magnitude higher than the average velocity 0 ± 5 m/s. The time delay space coupling $\mathbf{g}(t)$ is taken to be the identity matrix, as all the height measurements are assumed to be known with equal temporal precision throughout the observation window.

The time delay was selected to be $\tau = 10 \Delta t = 0.1$ h, in order to maintain a balance between numerical stability and the common criterion of nonlinear independence between the components of $\mathbf{S}(\mathbf{x}(t))$. The first minimum of the average mutual information was also calculated to be $\tau \approx 30 \Delta t$ using the method given by Abarbanel (1996). This is reasonably close to our choice, and the results did not change if its value was shifted by a few Δt.

9.7.1 Choosing D_M

The state was estimated by integrating the coupled differential equations Eq. (9.13) from $t = 0$ to $T = 5\,500$ hours Δt with various $D_M = \{1, 6, 8, 10\}$. The coupling terms were then switched off at $t = t_{final}$ to generate predictions until $t = 500$ hour. Short and long term synchronization error Eq. (9.14) trajectories $SE(t)$ are plotted in Fig. 9.11 for various D_M. Choosing $D_M = \{1, 6\}$ yields a synchronization error that remains around its initial value of 0.005 until the end of the five hour observation window. After the coupling is switched off, the error rises

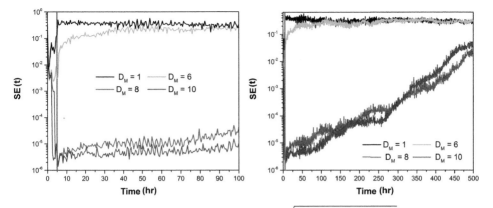

Figure 9.11 Synchronization error $SE(t_n) = \sqrt{\frac{1}{L} |\mathbf{x}^s(t_n) - \mathbf{y}^s(t_n)|^2}$, computed with $D_M = \{1, 6, 8, 10\}$, $G_h \Delta t = 1.5$, $G_u \Delta t = g_v \Delta t = 0.5$, and $\tau = 10 \Delta t = 0.1\,h$. Assimilation is performed for $t \le 5$ hr. **Left Panel** The couplings are then switched off and predictions are generated using the original dynamical equations (Eq. (9.15)) until $t = 100$ hr. In the prediction window ($t \ge 5$), the error in the trajectories grows roughly with the largest Lyapunov exponent of the system $\lambda_{max} \approx 1/3$ 1h. Synchronization is evident when $D_M = \{8, 10\}$ and not for $D_M = \{1, 6\}$, suggesting that accurate predictions will be obtained with $D_M = \{8, 10\}$. **Right Panel** The same calculation, but extended to $t = 500$ hr.

very rapidly until stabilizing around 0.1 for the remainder of the prediction window. By contrast, for $D_M = \{8, 10\}$ the synchronization error falls steeply to order 10^{-6} within the observation window. It then subsequently rises as $e^{\lambda_{max}(t-t_{final})}$, where $\lambda_{max} \approx 1/3$1h agrees with the largest Lyapunov exponent calculated for this flow. This exponential rate of growth is particularly evident in the long trajectory displayed in the Right Panel of Fig. 9.11. Since $D_M \ge 8$ produces error values several orders of magnitude smaller than those obtained with $D_M \le 6$, we expect the state estimates $\mathbf{x}(t_{final})$ obtained with $D_M \ge 8$ to be quite accurate when compared with the estimates for $D_M \le 6$. These estimates are now evaluated as they would be in a true experiment, by comparing predictions on the observed heights with additional data. In the prediction window $t > 5$ no information about the new measurements is passed back to the model. In Fig. 9.2 the known (black), estimated (red), and predicted (blue) height trajectories are shown for an arbitrarily selected grid point $h^{(6,4)}(t)$. Short and long term prediction trajectories computed with $D_M = 6$ are displayed in Fig. 9.11 upper panels, respectively. Corresponding results for $D_M = 8$ are shown in the lower panels. As anticipated, the predictions for $D_M = 8$ are clearly superior to those obtained with $D_M = 6$. The failure of predictions obtained with $D_M = 6$ is a result of poor estimates of the unobserved states (i.e. fluid velocities) at $t = t_{final}$. Although in an actual experiment we would not be able to verify this statement directly, we may do so here within a

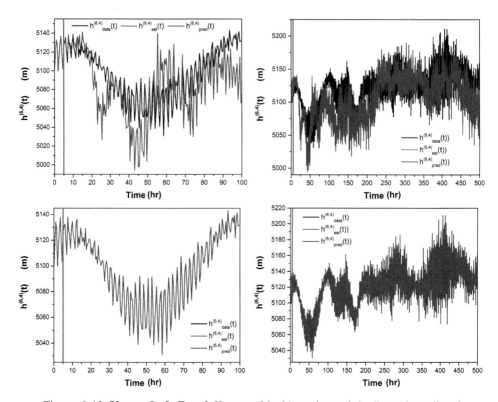

Figure 9.12 **Upper Left Panel** Known (black), estimated (red), and predicted (blue) for the observed height values $h^{(6,4)}(t)$ at grid point $(6, 4)$ for $D_M = 6$. Observations are for $0 \leq t \leq 5$ hr. Predictions are for $5 \leq t \leq 100$ hr. **Upper Right Panel** The same calculation for $D_M = 6$ for a longer prediction window $5 \leq t \leq 500$ hr. **Lower Left Panel** The same calculation except $D_M = 8$. Prediction window is $5 \leq t \leq 100$ hr. **Lower Right Panel** The same calculation except $D_M = 8$. Prediction window is $5 \leq t \leq 500$ hr.

twin experiment. Velocity profiles $u^{(6,4)}(t)$ displaying short and long time comparisons between the known (black), estimated (red), and predicted (blue) values are given in Fig. 9.3 for $D_M = 6$ in the upper panels, and for $D_M = 8$ in the lower panels. We find the situation is indeed as anticipated; the estimates and predictions are quite unacceptable for $D_M = 6$, whereas for $D_M = 8$ they are highly accurate. The same striking improvement in predictive accuracy was obtained for the other velocity component $v^{(6,4)}(t)$. These results are plotted in Fig. 9.14.

Predictions were also calculated for $D_M = 1$ and $D_M = 10$, but these results are not shown. They agree with the synchronization error calculations in Fig. 9.10, in that the predictions generated with $D_M = 10$ are just as accurate as those for $D_M = 8$. Likewise, predictions with $D_M = 1$ (i.e. simple nudging) are very poor, in accordance with Whartenby et al. (2013).

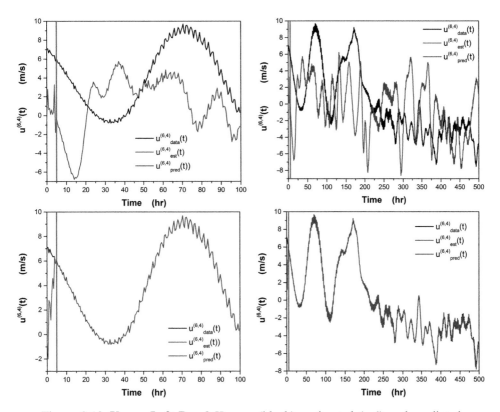

Figure 9.13 Upper Left Panel Known (black), estimated (red) and predicted (blue) for the observed x-velocity values $u^{(6,4)}(t)$ at grid point (6, 4) for $D_M = 6$. Observations are for $0 \le t \le 5$ hr. Predictions are for $5 \le t \le 100$ hr. **Upper Right Panel** The same calculation for $D_M = 6$ for a longer prediction window $5 \le t \le 500$ hr. **Lower Left Panel** The same calculation except $D_M = 8$. Prediction window is $5 \le t \le 100$ hr. **Lower Right Panel** The same calculation except $D_M = 8$. Prediction window is $5 \le t \le 500$ hr.

9.7.2 Further Reducing the Number of Measurements

In addition, until now we have conveniently chosen to observe the height field at all $L = N_\Delta^2 = 256$ grid locations. We now attempt to reduce L even further, by repeating the analysis with $L = 252$ and $L = 248$ height measurements, chosen at arbitrary grid points. From the results displayed in the Upper Left Panel of Fig. 9.15, it is evident that for $L = 252$ rapid and accurate synchronization is still achieved, while for $L = 248$ it is not. In addition, the known (black), estimated (red), and predicted values (blue) for $h^{(6,4)}(t)$ are shown in the other panels of Fig. 9.15 for $L = 248$ and $L = 252$, respectively. Results for the unobserved velocity fields agree as well, though these results are not shown.

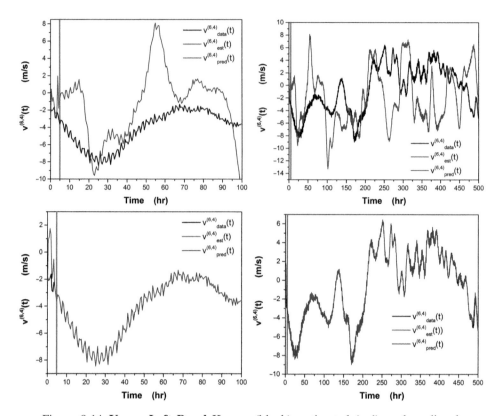

Figure 9.14 **Upper Left Panel** Known (black), estimated (red), and predicted (blue) for the observed y-velocity values $v^{(6,4)}(t)$ at grid point (6, 4) for $D_M = 6$. Observations are for $0 \leq t \leq 5$ hr. Predictions are for $5 \leq t \leq 100$ hr. **Upper Right Panel** The same calculation for $D_M = 6$ for a longer prediction window $5 \leq t \leq 500$ hr. **Lower Left Panel** The same calculation except $D_M = 8$. Prediction window is $5 \leq t \leq 100$ hr. **Lower Right Panel** The same calculation except $D_M = 8$. Prediction window is $5 \leq t \leq 500$ hr.

Thus, even with time delays, it may not be possible to significantly reduce the number of required height measurements. Additional refinement of the parameters $G(t)$, $g(t)$, D_M, and τ may further reduce this constraint, for instance by allowing $G(t)$ to be non-diagonal.

9.7.3 Noise in the Observations

We now repeat the above calculations for $L = 252$ with Gaussian noise $N(0, \sigma)$ added to the height observations. A comparison is shown in Fig. 9.15 for $\sigma = \{0.2, 0.5\}$ and $D_M = \{8, 10\}$. The synchronization error still falls rapidly within the observation window, although not to $O(10^{-5})$, as in the noiseless case. In the

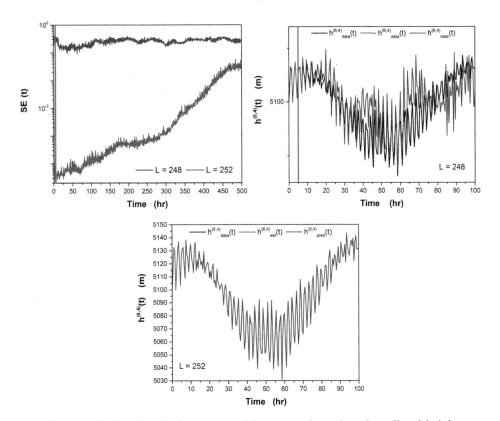

Figure 9.15 Synchronization error and known, estimated, and predicted height values for $L = 248$ height measurements at each observation time and for $L = 252$ height measurements at each observation time. **Upper Left Panel** $SE(t)$ for $L = 248$ and $L = 252$ over $0 \leq t \leq 5$ h in the observation window, and $5 \leq t \leq 500$ h after the couplings are removed. **Upper Right Panel** Known (black), estimated (red), and predicted (blue) values of the height $h^{(6,4)}(t)$ at gridpoint (6,4) for $0 \leq t \leq 100$ h for $L = 248$. **Lower Panel** Known (black), estimated (red), and predicted (blue) values of the height $h^{(6,4)}(t)$ at gridpoint (6,4) for $0 \leq t \leq 100$ h for $L = 252$. This shows the rather sharp transition between bad predictions ($L = 248$) and good predictions ($L = 252$).

prediction window, it rises in an exponential manner as expected. These results were included to show that the method appears to be relatively robust to small errors in the observations.

9.7.4 Using Drifter Data

Another quite important source of observations about ocean flows is being provided by position measurements $\mathbf{r}(t)$ of Lagrangian drifters (Mariano et al. (2002); An et al. (2017)).

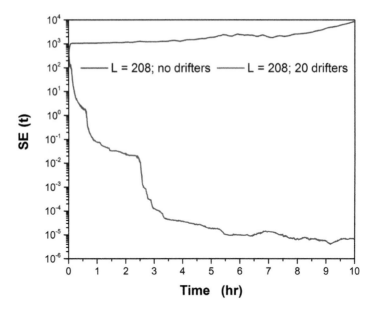

Figure 9.16 The effect of noise levels in the initial condition for the solution of the model equations Eq. (9.15) on $SE(t)$. We show the results for $D_M = 8$ and 10 for added Gaussian noise $N(0, \sigma)$ with $\sigma = 0.2$ and 0.5. For this range of noise levels added to the initial condition for generating the data in our twin experiments, we see that the detailed values of $SE(t)$ change. In the case of both $D_M = 8$ and $D_M = 10$, $SE(t)$ still becomes quite small in the observation window $0 \leq t \leq 5$ h, suggesting that predictions for $t \geq 5$ will remain robustly accurate.

Such observations have been shown to be a good supplement to the traditional observations made on a fixed grid (Kuznetsov et al. (2003)), and they can also be used to estimate an Eulerian velocity field (Molcard et al. (2003); Piterbarg (2008); Salman et al. (2006)). In this section, we combine the time delay method with a data set from drifter measurements to show that they can provide accurate estimates for the grid state variables $\{h(\mathbf{r}^{(i,j)}, t), u(\mathbf{r}^{(i,j)}, t), v(\mathbf{r}^{(i,j)}, t)\}$, without much additional effort.

We monitor the positions of N_D drifters deployed at randomly chosen initial grid locations and afterwards allowed to move freely to provide spatially continuous measurements between grid points. The dynamics of drifters are described as two-dimensional fluid parcel motion on the surface of the water layer, which are determined by the Lagrangian equations

$$\frac{d\mathbf{r}^{(n)}(t)}{dt} = \mathbf{u}(\mathbf{r}^{(n)}(t), t)$$

where $\mathbf{r}^{(n)}(t)$ is the position of the nth; $n = 1, 2, \ldots, N_D$ drifter. This equation was simulated by linear interpolation of the discrete velocity fields (Press et al. (2007);

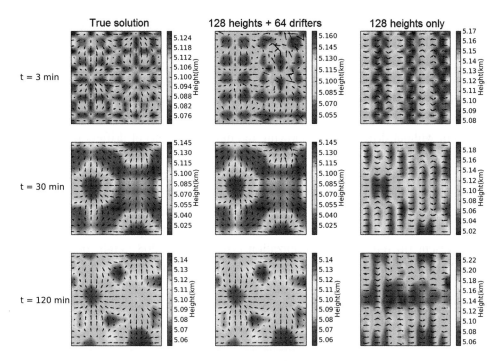

Figure 9.17 Comparison of the estimated and predicted fields $\{h(t), u(t), v(t)\}$ between the data (**Left Column**) and analyses, run with observations of 128 height variables, both with (**Center Column**) and without drifters (**Right Column**). Snapshots are taken 3 min (**Upper Row**) into the assimilation window, at 30 min the end of the assimilation window (**Center Row**), and 90 min into the prediction window (**Bottom Row**).

Thomson and Emery (2014)). Hybrid measurements are incorporated into the time delay nudging method by combining the grid variables and the collective drifter positions

$$\mathbf{R}^T(t) = \{[\mathbf{r}^{(1)}(t)]^T, [\mathbf{r}^{(2)}(t)]^T, \ldots, [\mathbf{r}^{(N_D)}(t)]^T\}$$

into a single hybrid state vector. The corresponding time delayed vectors are $\mathbf{Y}_{drifter}(t) = \{\mathbf{Y}_{grid}(t), \mathbf{Y}_{drifter}(t)\}$ and time delayed state vectors $\mathbf{S}_{drifter}(t) = \{\mathbf{S}_{grid}(t), \mathbf{S}_{drifter}(t)\}$, respectively, where

$$\mathbf{Y}^T_{drifter}(t) = \{\mathbf{R}^T_{data}(t), \mathbf{R}^T_{data}(t + \tau), \ldots, \mathbf{R}^T_{data}(t + \tau(D_M - 1))\}$$

$$\mathbf{S}^T_{drifter}(t) = \{\mathbf{R}^T_{model}(t), \mathbf{R}^T_{model}(t + \tau), \ldots, \mathbf{R}^T_{model}(t + \tau(D_M - 1))\}. \quad (9.16)$$

In contrast to the previous results, where the initial conditions for the grid variables were taken to differ in both phase and frequency between the true solution and the estimate, here the initial conditions only vary in amplitude. That is, the initial conditions of the data $\psi_{data}(\mathbf{r}^{(i,j)}, t_0)$ and $h_{data}(\mathbf{r}^{(i,j)}, t_0)$ and of the model $\psi_{model}(\mathbf{r}^{(i,j)}, t_0)$ and $h_{model}(\mathbf{r}^{(i,j)}, t_0)$ are related by $\psi_{data}(\mathbf{r}^{(i,j)}, t_0) = C_0 \psi_{model}(\mathbf{r}^{(i,j)}, t_0)$ and $h_{data}(\mathbf{r}^{(i,j)}, t_0) = C_0 h_{model}(\mathbf{r}^{(i,j)}, t_0)$. We choose $C_0 = 1.0 + 0.1\,\eta$, with η selected from a uniform distribution in the interval $[-1, 1]$. The velocity fields are found as above, using $\psi(\mathbf{r}, t_0)$ as a stream function. This was done in order to improve the results, as we found that the drifter results were more sensitive to the choice of initial condition than the results from the previous section, without drifters. Plots showing the initial positions of drifters for the two cases considered below ($N_D = 20$ and $N_D = 64$) are shown in Fig. 9.16. They were also deactivated when they reached the boundary of the grid, so the number of operational drifters decreases throughout the estimation window.

In Fig. 9.18, we show the synchronization error of observed quantities for $D_M = 8$, keeping all other parameters the same as in the previous calculations. We present (in red) the synchronization error for L = 208 height observations and $N_D = 20$ drifter observations, and we show (in blue) the same synchronization error when $L = 208$ and $N_D = 0$ drifters are deployed. With $L = 208$, namely, observing 27 percent of the heights and 20 drifters, the synchronization

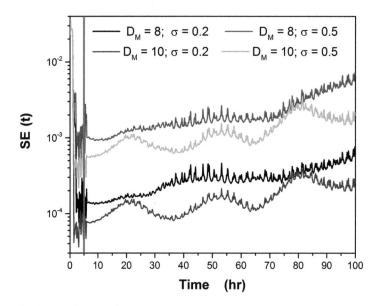

Figure 9.18 Synchronization error $SE(t)$ for our standard twin experiment when we utilize drifter information, and then when we do not utilize drifter information. When the number of observations of height is $L = 208$, we see that without drifter information (blue line) there is no synchronization and correspondingly inaccurate predictions (not shown). When information from 20 Lagrangian drifters is added during data assimilation using time delay nudging, $SE(t)$ decreases very rapidly (red line) indicating predictions will be very accurate (also not shown). The efficacy of even small numbers of drifters is clear in this example.

error converges to a small value within the five-hour observation window. Without drifters, the estimation fails.

By increasing the number of drifters to $N_D = 64$ within a 30 minute observation window, synchronization can be achieved with $L = 128$ height observations. Snapshots of the fields at different times throughout the estimation and prediction window are shown in Fig. 9.18 for comparison.

One rather striking result was that when L = 128 heights were observed over six hours without drifters, the synchronization error remained large. The homogeneous addition of $N_D = 64$ drifters reduced the synchronization error nearly to zero.

This underlines the utility of adding time delay coordinates to the observed drifter locations to extract further information useful to determining the state of the model for an initial state for predictions (Molcard et al. (2003); Piterbarg (2008); Salman et al. (2006)).

It is clear from these preliminary results that drifter data can be useful for improving the observability of the system, and that the time delay method provides a way to incorporate this information into the analysis.

9.8 Summing It Up

The transfer of information from measurements of a chaotic dynamical system to a quantitative model of the system is impeded when the number of measurements at each measurement time is below an approximate threshold L_c, which can be established in a twin experiment. Whartenby et al. (2013) previously showed that for a nonlinear model for shallow water flow, a standard nudging technique given by Eq. (9.7) requires direct observation of roughly 70% of the dynamical variables $\{h(\mathbf{r}, t), u(\mathbf{r}, t), v(\mathbf{r}, t)\}$ at each measurement time to synchronize the model output with the observations.

We have demonstrated how information in the time delays of the observations may be used to reduce this requirement to about 30%, in which only the height fields need be observed. Moreover, it appears L_c can be even further reduced by adding positional information from drifters, which interpolate the height field at locations between grid points (An et al. (2017)).

Although all this has been done on a simplified model of shallow water flow, implemented with only $D = 3N_\Delta^2 = 768$ degrees of freedom, on a β-plane driven by surface winds, the process can be used to analyze increasingly realistic and complex models of coupled earth systems.

Since the successful analysis of simulated data is probably a prerequisite for success with real data, when the model is wrong, which it will be in practice, this methodology provides insight as to whether the model is at fault, or whether more observations are needed.

10

Unfinished Business

There is always room for a chapter with this title. It serves as a reminder that there is likely to be more to do, whatever one has already done.

Here are a few items to consider:

- Explore in some depth the Euler-Lagrange Equations for the DA standard model, Eq. (4.18), and for deepest learning, Eq. (9.19), to determine the path(s) with minimum action levels. These are two point boundary value problems where the canonical momenta vanish at initial times and layers (t_0, l_0) and final times and layers (t_F, l_F).

- The dynamical systems addressed in this book have been completely classical. $\hbar = 0$ is far from encompassing many or most of the interesting questions of the twenty-first century. How shall we extend all the approaches discussed in the Chapters before this one in a quantum mechanical context? One start at that is found in Abarbanel (2011), and an application to neutrino flavor evolution is available. In quantum theory path integrals are natural vehicles for the discussion of data assimilation (Feynman and Hibbs (1965)), so starting out should be rather easy.

  Two new features, discussed for nearly a century, are the discrete outcomes of some quantum mechanical measurements and the need to work with complex probability amplitudes rather than real probabilities.

- In our consideration of examples from neurobiology we did not stray from data assimilation of individual, isolated neurons. The object of one's focus should be expanded to functional networks of neurons. This also requires experiments where correlations among neurons are measured and taking into account their synaptic and gap junction connectivity.

- The machine learning examples discussed here have simple network architectures, which the original investigators (Goodfellow et al. (2016); LeCun et al.

(2015)) presented as representing the manner in which complex mammalian brains operate.

This has long been recognized as not the way biophysical nervous systems actually work, and it may be fruitful to intensely analyze the way simple, smaller functional networks operate to acquire guidance at the level of order 10^{3-4} neurons and set a path toward the 10^{11} neurons we might aspire to understand. Along the way, by analyzing such small functional nervous systems, we will certainly both gain insights into architectures to achieve machine learning tasks and systematically gain insights on how to build quantitative models of richer nervous systems.

- One should expand the class of scientific and technical problems to which data assimilation and machine learning, as seen from a path integral point of view, can be successfully applied. This monograph drew examples from neurobiology and relatively simple geophysical problems. The limitation to these areas has not been fundamental at all, but it comes from the author's inability to handle more than a couple topics at a given time.

Given the breadth of inquiries in science and technology and the concomitant increase in computational capability, this might properly be identified as 'unfinishable business.'

There are inevitably 'surprises' in using a general formulation, such as in this book, to specific problems in science and technology. Measurement error as required for an action $A(\mathbf{X}) = -\log[P(\mathbf{X})] = \mathbf{Model\ Error\ +\ Measurement\ Error}$ requires understanding the performance of specific instruments. Model error requires a model relevant for the issues in the problem and some way to represent errors in the model.

Exercising the general formulation on problems of many sizes and complexity will tell us quantitative answers to questions such as "How many measurements are required at each observation time?" (see Chapter 9, for example). This will certainly reveal innovative methods addressing all the issues covered within the previous chapters.

I suggested early on in this book that everyone who is engaged in building and testing nonlinear dynamical models for observed phenomena would find material relevant to their scientific inquiries in the discussions covered here. That continues to be my viewpoint.

Bibliography

Abarbanel, H., Creveling, D., Farsian, R., and Kostuk, M. (2009). Dynamical state and parameter estimation. *SIAM Journal on Applied Dynamical Systems*, 8(4):1341–1381.

Abarbanel, H. D. I. (1996). *The Analysis of Observed Chaotic Data*. Springer-Verlag, New York.

Abarbanel, H. D. I. (2009). Effective actions for statistical data assimilation. *Physics Letters A*, 373(44):4044–4048.

Abarbanel, H. D. I. (2011). Private communication to Lu Sham.

Abarbanel, H. D. I. (2013). *Predicting the Future: Completing Models of Observed Complex Systems*. Understanding Complex Systems. Springer-Verlag, New York.

Abarbanel, H. D. I., Rozdeba, P. J., and Shirman, S. (2018). Machine learning; deepest learning as statistical data assimilation problems. *Neural Computation*, 30:2025–2055.

Abarbanel, H. D. I., Rul'kov, N. F., and Sushchik, M. M. (1996). The auxiliary systems approach to generalized synchronization of chaos. *Physical Review E*, 53:4528–4535.

Aeyels, D. (1981a). Generic observability of differentiable systems. *SIAM Journal on Control and Optimization*, 19:595–603.

Aeyels, D. (1981b). On the number of samples necessary to achieve observability. *Systems & Control Letters*, 1:92–94.

An, Z. (2019). *Data Assimilation by Reconstructing Time-Series Observations*. PhD thesis, University of California San Diego. Accessed at https://escholarship.org/uc/item/42k273gk.

An, Z., Rey, D., Ye, J., and Abarbanel, H. D. I. (2017). Estimating the state of a geophysical system with sparse observations: time delay methods to achieve accurate initial states for prediction. *Nonlinear Processes in Geophysics*, 24:9–22.

Anthes, R. (1974). Data assimilation and initialization of hurricane prediction models. *Journal of the Atmospheric Sciences*, Volume 31, pp 702–719.

Arnol'd, V. I. (1989). *Mathematical Methods of Classical Mechanics*. Graduate Texts in Mathematics, vol. 60. 2nd ed. Springer-Verlag, New York.

Bennett, A. F. (1992). *Inverse Methods in Physical Oceanography*. Cambridge University Press.

Betancourt, M. (2017). A conceptual introduction to Hamiltonian Monte Carlo. *arXiv:1701.02434v2*.

Betts, J. T. (2010). *Practical Methods for Optimal Control and Estimation Using Nonlinear Programming, Second Edition*. Society for Industrial and Applied Mathematics (SIAM), Philadelphia. *arXiv:1701.02434v2*.

Blanes, S., Casas, F., and Sanz-Serna, J. M. (2014). Numerical integrators for the Hybrid Monte Carlo method. *SIAM Journal on Scientific Computing*, 36(4):A1556–1580.

Blum, J. (2010). Data assimilation for geophysical problems: variational and sequential techniques. University of Nice Sophia Antipolis, France. Accessed at www.math.univ-toulouse.fr/ baehr/meteo_SMAI_DSNA/Pres/Pres_Blum.pdf

Blum, J., Le Dimet, F.-X., and Navon, I. M. (2009). Chapter in computational methods for the atmosphere and the oceans. In *Volume 14: Special Volume of Handbook of Numerical Analysis*. R. Temam and J. Tribbia, eds. Elsevier Science Ltd, New York.

Byrd, R. H., Lu, P., and Nocedal, J. (1995). A limited memory algorithm for bound constrained optimization. *SIAM Journal on Scientific and Statistical Computing*, 16:1190–1208.

Channell, P. and Scovel, J. (1990). Symplectic integration of Hamiltonian systems. *Nonlinearity*, 3:231–259.

Chua, B. S. and Bennett, A. F. (2001). An inverse ocean modeling system. *Ocean Modelling*, 3:137–165.

Chyba, M., Hairer, H., and Vilmart, G. (2009). The role of symplectic methods in optimal control. *Optimal Control Applications and Methods*, 30:367–382.

Cox, H. (1964). On the estimation of state variables and parameters for noisy dynamic systems. *IEEE Transactions on Automatic Control*, 9:5–12.

Creutz, M. (1988). Global Monte Carlo algorithms for many-fermion systems. *Physical Review D*, 38(4):1228.

Creveling, D. (2008). *Parameter and State Estimation in Nonlinear Dynamical Systems*. PhD thesis, University of California San Diego.

Dochain, D. (2003). State and parameter estimation in chemical and biochemical processes: a tutorial. *Journal of Process Control*, 13(8):801–818.

Duane, S., Kennedy, A. D., Pendleton, B. J., and Roweth, D. (1987). Hybrid Monte Carlo. *Physics Letter B*, 195(2):216–222.

Durr, D. and Bach, A. (1978). The Onsager–Machlup function as Lagrangian for the most probable path of a diffusion process. *Communications in Mathematical Physics*, 60:153–170.

Durstewitz, D. (2017). *Advanced Data Analysis in Neuroscience: Integrating Statistical and Computational Models*. Bernstein Series in Computational Neuroscience. Springer International Publishing, Basel, Switzerland.

Eckmann, J.-P. and Ruelle, D. (1985). Ergodic theory of chaos and strange attractors. *Reviews of Modern Physics*, 57:617–656.

Eldridge, M. J. (2016). *Use of Data Assimilation to Determine Features of Neuron Structure and Connectivity*. PhD thesis, University of California San Diego. Accessed at https://escholarship.org/uc/item/7st8j1r8.

Elman, J. L. (1990). Finding structure in time. *Cognitive Science*, 14:179–211.

Evensen, G. (2009). *Data Assimilation: The Ensemble Kalman Filter*. Springer-Verlag, New York.

Fang, Z., Wong, A. S., Hao, K., Ty, A. J. A., and Abarbanel, H. D. I. (2020). Precision annealing Monte Carlo methods for statistical data assimilation and machine learning. *Physical Review Research*, 2:013050.

Fano, R. M. (1961). *Transmission of Information; A Statistical Theory of Communication*. Massachusetts Institute of Technology Press.

Feller, W. (2008). *An Introduction to Probability Theory and Its Applications*, vol. 2. John Wiley & Sons, Hoboken, NJ.

Feynman, R. P. and Hibbs, A. R. (1965). *Quantum Mechanics and Path Integrals*. McGraw-Hill, New York.

Frank, R. J., Davey, N., and Hunt, S. P. (2001). Time series prediction and neural networks. *Journal of Intelligent and Robotic Systems*, 31:91–103.

Garey, M. R. and Johnson, D. S. (1990). *Computers and Intractability; A Guide to the Theory of NP-Completeness*. W. H. Freeman & Co., New York. ISBN:0716710455; see also: www.britannica.com/science/NP-complete-problem.

Gelfand, I. M. and Fomin, S. V. (1963). *Calculus of Variations*. Dover Publications, Mineola, NY.

Gelfand, I. M. and Yaglom, A. M. (1960). Integration in functional spaces and its applications in quantum physics. *Journal of Mathematical Physics*, 1(1):48–69.

Ghahraman, Z. (2015). Probabilistic machine learning and artificial intelligence. *Nature*, 521:452–459. doi: https://doi.org/10.1038/nature14541.

Ghil, M. and Malanotte-Rizzoli, P. (1991). Data assimilation in meteorology and oceanography. *Advances in Geophysics*, 33:141–266.

Girolami, M. and Calderhead, B. (2011). Riemann manifold Langevin and Hamiltonian Monte Carlo methods. *Journal of the Royal Statistical Society B*, 73:123–214.

Goldstein, H., Poole, C. P., and Safko, J. L. (2002). *Classical Mechanics*. 3rd ed. Pearson.

Goodfellow, I., Bengio, Y., and Courville, A. (2016). *Deep Learning*. Massachusetts Institute of Technology Press, Cambridge, MA; London. Accessed at www.deeplearningbook.org.

Greengard, L. and Rokhlin, V. (1991). On the numerical solution of two-point boundary value problems. *Communications on Pure and Applied Mathematics*, 44(4):419–452.

Gubernatis, J. E. (2005). Marshall Rosenbluth and the Metropolis algorithm. *Physics of Plasmas*, 12:057303.

Guckenheimer, J. and Oliva, R. A. (2002). Chaos in the Hodgkin-Huxley model. *SIAM Journal on Applied Dynamical Systems*, 1:105–114.

Gulshan, V., Peng, L., Coram, M., Stumpe, M. C., Wu, D., Narayanaswamy, A., Venugopalan, S., Widner, K., Madams, T., Cuardos, J., Kim, R., Raman, R., Nelson, P. C., Mega, J. L., and Webster, D. R. (2016). Development and validation of a deep learning algorithm for detection of diabetic retinopathy in retinal fundus photographs. *Journal of the American Medical Association*, 316:2402–2410. doi: https://doi.org/10.1001/jama.2016.17216.

Gupta, K. (2008). Accessed at www.britannica.com/science/NP-complete-problem.

Hairer, E., Wanner, G., and Lubich, C. (2006). *Geometric Numerical Integration: Structure-Preserving Algorithms for Ordinary Differential Equations*. Springer Series in Computational Mathematics, vol. 31, 2nd ed. Springer-Verlag, Berlin and Heidelberg.

Hairer, E. and Zbinden, C. J. (2012). Conjugate symplectic b-series. In *Numerical Analysis and Applied Mathematics INCAAM*, pages 23–26. AIP Conference Proceedings. https://doi.org/10.1063/1.4756053

Hastings, W. K. (1970). Monte Carlo sampling methods using Markov chains and their applications. *Biometrika*, 57:97–109.

Haugh, M. (2017). MCMC and Bayesian modeling; lecture notes. Technical report. IEOR E4703: Monte-Carlo Simulation, Columbia University. Accessed at www.columbia.edu/~mh2078/MonteCarlo.html

Hochberg, D., Molina-Paris, C., Perez-Mercader, J., and Visser, M. (1999). Effective action for stochastic partial differential equations. *Physical Review E*, 60(6):6343.

Hodgkin, A. L. and Huxley, A. F. (1952). A quantitative description of membrane current and its application to conduction and excitation in nerve. *The Journal of Physiology*, 117(4):500–544.

Horváth, A. and Manini, D. (2008). Parameter estimation of kinetic rates in stochastic reaction networks by the em method. In *International Conference on BioMedical Engineering and Informatics, 2008. BMEI 2008*. vol. 1, pages 713–717. IEEE.

Johnston, D. and Wu, S. M.-S. (1995). *Foundations of Cellular Neurophysiology*. Bradford Books, Massachusetts Institute of Technology Press.

Jordan, M. (1986). Attractor dynamics and parallelism in a connectionist sequential machine. In *Proceedings of the Eighth Conference of the Cognitive Science Society*, pages 531–546. Cognitive Science Society.

Jouvet, B. and Phythian, R. (1979). Quantum aspects of classical and statistical fields. *Physical Review A*, 19(3):1350.

Kadakia, N. (2016). Hybrid monte carlo with chaotic mixing. *arXiv:1604.07343*.

Kadakia, N. (2017). *The Dynamics of Nonlinear Inference*. PhD thesis, University of California San Diego.

Kadakia, N., Rey, D., Ye, J., and Abarbanel, H. D. I. (2017). Symplectic structure of statistical variational data assimilation. *Quarterly Journal of the Royal Meteorological Society*, 143(703):756–771.

Kalnay, E. (2003). *Atmospheric Modeling, Data Assimilation, and Predictability*. Cambridge University Press.

Kantz, H. and Schreiber, T. (2004). *Nonlinear Time Series Analysis, 2nd ed*. Cambridge University Press.

Kirk, D. E. (1970). *Optimal Control Theory: An Introduction*. Dover Publications, Mineola, NY.

Kirkpatrick, S., Gelatt Jr, C. D., and Vecchi, M. P. (1983). Optimization by simulated annealing. *Science*, 220:671–680.

Knowlton, C. (2014). *Path Integral Techniques for Estimating Neural Network Connectivity*. PhD thesis, University of California San Diego. https://escholarship.org/uc/item/2jc5153h.

Kostuk, M. (2012). *Synchronization and Statistical Methods for the Data Assimilation of HVC Neuron Models*. PhD thesis, University of California San Diego.

Kostuk, M., Toth, B., Meliza, C., Margoliash, D., and Abarbanel, H. (2012). Dynamical estimation of neuron and network properties ii: path integral Monte Carlo methods. *Biological Cybernetics*, 106(3):155–167.

Kot, M. (2014). *A First Course in the Calculus of Variations*. American Mathematical Society, Providence, RI.

Kuznetsov, L., Ide, K., and Jones, C. K. R. T. (2003). A method for assimilation of Lagrangian data. *Monthly Weather Review*, 131:2247–2260.

Lall, S. and West, M. (2006). Discrete variational Hamiltonian mechanics. *Journal of Physics A: Mathematical and General*, 39(19):5509.

Laplace, P. (1986). Memoir of the probability of causes of events. *Statistical Science*, 1:365–378. Translation to English by S. M. Stigler.

Laplace, P. S. (1774). Memoir on the probability of causes of events. *Mathématique et de Physique, Tome Sixiéme*, pages 621–656.

LeCun, Y., Bengio, Y., and Hinton, G. (2015). Deep learning. *Nature*, 521:436–444.

Lei, B., Xu, G., Feng, M., Zou, Y., van der Heijden, F., de Ridder, D., and Tax, D. M. J. (2017). *Classification, Parameter Estimation and State Estimation: An Engineering Approach Using MATLAB 2nd Edition*. John Wiley and Sons, Hoboken, NJ.

Leok, M. and Zhang, J. (2011). Discrete Hamiltonian variational integrators. *IMA Journal of Numerical Analysis*, 31(4):1497–1532, doi: https://doi.org/10.1093/imanum/drq027.

Lewis, J. M. and Derber, J. C. (1985). The use of adjoint equations to solve a variational adjustment problem with advective constraints. *Tellus A*, 37A:309–322.

Liberzon, D. (2012). *Calculus of Variations and Optimal Control Theory*. Princeton University Press.

Lorenc, A. C. and Payne, T. (2007). 4D-Var and the butterfly effect: statistical four-dimensional data assimilation for a wide range of scales. *Quarterly Journal of the Royal Meteorological Society*, 133(624):607–614.

Lorenz, E. N. (1963). Deterministic nonperiodic flow. *Journal of the Atmospheric Sciences*, 20(2):130–141.

Lorenz, E. N. (2006). Predictability: a problem partly solved. In Palmer, T. and Hagedorn, R., eds., *Predictability of Weather and Climate*. Cambridge University Press.

Lorenz, E. N. and Emanuel, K. A. (1998). Optimal sites for supplementary weather observations: simulation with a small model. American Meteorological Society, MA.

Mañé, R. (1981). On the dimension of the compact invariant sets of certain nonlinear maps. In Rand, D. A. and Young, L.-S., eds. *Dynamical Systems and Turbulence, Lecture Notes in Mathematics*, vol. 898, pages 230–242.

Mangoubi, O. and Vishnoi, N. K. (2018). Dimensionally tight bounds for second-order Hamiltonian Monte Carlo. *Proceedings of the 32nd International Conference on Neural Information Processing Systems*, pages 6030–6040.

Mariano, A. J., Griffa, A., Zgkmen, T. M., and E. Zambianchi, E. (2002). Lagrangian analysis and predictability of coastal and ocean dynamics. *Journal of Atmospheric and Oceanic Technology*, 19:1114–1126.

Marsden, J. E. and West, M. (2001). Discrete mechanics and variational integrators. *Acta Numerica*, pages 357–514.

Metropolis, N., Rosenbluth, A. W., Rosenbluth, M. N., Teller, A. H., and Teller, E. (1953). Equation of State Calculations by Fast Computing Machines. *Journal of Chemical Physics*, 21:1087–1092.

Miller, K. (1970). Least squares methods for ill-posed problems with a prescribed bound. *SIAM Journal on Mathematical Analysis*, 1:52–74.

Miller, R. N., Ghil, M., and Gauthiez, F. (1994). Advanced data assimilation in strongly nonlinear dynamical systems. *Journal of the Atmospheric Sciences*, 51:1037–1056.

Mitre17 (2017). *Perspectives on Research in Artificial Intelligence and Artificial General Intelligence Relevant to DoD*. https://apps.dtic.mil/sti/citations/AD1024432.

Molcard, A., Piterbarg, L., A. Griffa, A., Özgökmen, T. M., and Mariano, A. J. (2003). Assimilation of drifter observations for the reconstruction of the eulerian circulation field. *Journal of Geophysical Research: Oceans*, 593:2156–2202.

Morone, U. I. (2016). *Predicting the Behavior of Dynamical and Biological Systems Using Asynchronous Data*. PhD thesis, University of California San Diego. Accessed at https://escholarship.org/uc/item/4s2045vj.

Murty, K. G. and Kabadi, S. N. (1987). Some np-complete problems in quadratic and nonlinear programming. *Mathematical Programming*, 39:117–129.

Neal, R. M. (1993). Probabilistic inference using Markov chain Monte Carlo methods. Technical report, Department of Computer Science, University of Toronto. Technical Report CRG-TR-93-1.

Neal, R. M. (2011). MCMC using Hamiltonian dynamics. In Brooks, S., Gelman, A., Jones, G., and Meng, X.-L., eds. *Handbook of Markov Chain Monte Carlo*, chapter 5, pages 113–162. Chapman and Hall/CRC, Boca Raton, FL.

Nogaret, A., Meliza, C. D., Margoiliash, D., and Abarbanel (2016). Automatic construction of predictive neuron models through large scale assimilation of electrophysiological data. *Scientific Reports*, 6:32749. doi: https://doi.org/10.1038/srep32749.

Onsager, L. and Machlup, S. (1953). Fluctuations and irreversible processes. *Physical Review*, 91(6):1505.

Oseledec, V. I. (1968). A multiplicative ergodic theorem. Lyapunov characteristic numbers for dynamical systems. *Trudy Moskovskogo Matematicheskogo Obshchestva*, 19:197–231.

Ott, E. (2019). Using machine learning for prediction of large complex, spatially extended systems. Workshop on Dynamical Methods in Data-Based Exploration of Complex Systems. Max-Planck Institute for the Physics of Complex Systems, Dresden, Germany.

Palatella, L., Carrassi, A., and Trevisan, A. (2013). Lyapunov vectors and assimilation in the unstable subspace: theory and applications. *Journal of Physics A: Mathematical and Theoretical*, 46:25.

Parlitz, U., Junge, L., and Kocarev, L. (1996). Synchronization-based parameter estimation from time series. *Physical Review E*, 54(6):6253.

Pathak, J., Hunt, B., Girvan, M., Lu, Z., and Ott, E. (2018). Model-free prediction of large spatiotemporally chaotic systems from data: a reservoir computing approach. *Physical Review Letters*, 120:024102.

Pazó, D., Carrassi, A., and López, J. M. (2016). Data assimilation by delay-coordinate nudging. *Quarterly Journal of the Royal Meteorological Society*, 142:1290–1299. https://doi.org/10.1002/qj.2732.

Pecora, L. M. and Carroll, T. L. (1990). Synchronization in chaotic systems. *Physical Review Letters*, 64:821–825.

Pedlosky, J. (1986). *Geophysical Fluid Dynamics, 2nd Edition*. Springer Verlag. ISBN 0-387-96387-1.

Peng, L. (2018). Assessing cardiovascular risk factors with computer vision. Technical report, Google Brain Team. https://ai.googleblog.com/2018/02/.

Pires, C., Vautard, R., and Talagrand, O. (1996). On extending the limits of variational assimilation in nonlinear chaotic systems. *Tellus A*, 48(1):96–121.

Piterbarg, L. I. (2008). Optimal estimation of Eulerian velocity field given Lagrangian observations. *Applied Mathematical Modeling*, 32:2133–2148.

Pontryagin, L. S. (1959). Optimal control processes. *Uspekhi Fizicheskikh Nauk*, 14(1): 3–20.

Press, W. H., Teukolsky, S. A., Vetterling, W., and Flannery, B. (2007). *Numerical Recipes, The Art of Scientific Computing, 3rd Edition*. Cambridge University Press.

Proust, M. (1913). *À la recherche du temps perdu; Remembrance of Things Past*. Grasset and Gallimard. Translators: C. K. Scott Moncrieff, Stephen Hudson, Terence Kilmartin, Lydia Davis, and James Grieve; 4215 pages; 7 volumes; 1913–1927; French. 1922–1931; English.

Quinn, J. C. (2010). *A Path Integral Approach to Data Assimilation in Stochastic Nonlinear Systems*. PhD thesis, University of California San Diego.

Quinn, J. C. and Abarbanel, H. D. I. (2010). State and parameter estimation using Monte Carlo evaluation of path integrals. *Quarterly Journal of the Royal Meteorological Society*, 136(652):1855–1867.

Rabier, F., Järvinen, H., Klinker, E., Mahfouf, J. F., and Simmons, A. (1994). The ECMWF operational implementation of four-dimensional variational assimilation. Part I: Experimental results with simplified physics. *Quarterly Journal of the Royal Meteorological Society*, pages 1143–1170.

Reich, S. and Cotter, C. (2015). *Probabilistic Forecasting and Bayesian Data Assimilation*. Cambridge University Press.

Rey, D. (2017). *Chaos, Observability and Symplectic Structure in Optimal Estimation*. PhD thesis, University of California San Diego. https://escholarship.org/uc/item/3049w2dh.

Rey, D., Eldridge, M., Kostuk, M., Abarbanel, H. D. I., Schumann-Bischoff, J., and Parlitz, U. (2014a). Accurate state and parameter estimation in nonlinear systems with sparse observations. *Physics Letters A*, 378(11):869–873.

Rey, D., Eldridge, M., Morone, U., Abarbanel, H. D. I., Parlitz, U., and Schumann-Bischoff, J. (2014b). Using waveform information in nonlinear data assimilation. *Physical Review E*, 90:062916.

Rissanen, J. (1989). *Stochastic Complexity in Statistical Inquiry Theory*. World Scientific Publishing Co., Inc., River Edge, NJ.

Rozdeba, P. J. (2017). *Nonlinear Inference in Partially Observed Physical Systems and Deep Neural Networks*. PhD thesis, University of California San Diego. Accessed at https://escholarship.org/uc/item/3h22m81p.

Rozdeba, P. J. (2018). A Python package for state and parameter estimation in partially observed ode and neural network systems, using variational annealing. Accessed at https://github.com/paulrozdeba/varanneal.

Ruth, R. D. (1983). A canonical integration technique. *IEEE Transactions on Nuclear Science*, NS-30:2669–2671.

Sadourny, R. (1975). The dynamics of finite-difference models of the shallow-water equations. *Journal of the Atmospheric Sciences*, 32:680–689.

Salman, H., Kuznetsov, L., Jones, C. K. R. T., and Ide, K. (2006). A method for assimilating lagrangian data into a shallow-water-equation ocean model. *Monthly Weather Review*, 134:1081–1101.

Sanz-Serna, J. M. (1992). Symplectic integrators for Hamiltonian problems: an overview. *Acta Numerica*, 1:243–286.

Sanz-Serna, J. M. (2016). Symplectic Runge-Kutta schemes for adjoint equations, automatic differentiation, optimal control, and more. *SIAM Review*, 58:3–33.

Sauer, T. J., Yorke, J. A., and Casdagli, M. (1991). Embedology. *Journal of Statistical Physics*, 65:579–616.

Schwartz, M. D. (2014). *Quantum Field Theory and the Standard Model*. Cambridge University Press. ISBN 978-1-107-03473-0.

Senselab-Yale (2020). Accessed at https://senselab.med.yale.edu/.

Shirman, S. (2018). *Strategic Monte Carlo and Variational Methods in Statistical Data Assimilation for Nonlinear Dynamical Systems*. PhD thesis, University of California San Diego.

Sohl-Dickstein, J., Mudigonda, M., and DeWeese, M. R. (2016). Hamiltonian Monte Carlo without detailed balance. In *Proceedings of the 31st International Conference on Machine Learning, Beijing, China, 2014*. arXiv: 1409.5191v5.

Sterratt, D., Graham, B., Gillies, A., and Willshaw, D. (2011). *Principles of Computational Modelling in Neuroscience*. Cambridge University Press. ISBN: 978052 1877954.

Sushchik, M. M., Rul'kov, N. F., Tsimring, L. S., and Abarbanel, H. D. I. (1995). Generalized synchronization of chaos in directionally coupled chaotic systems. *Physical Review E*, 51:980–994.

Suwa, H. and Todo, S. (2010). Markov chain Monte Carlo method without detailed balance. *Physical Review Letters*, 105:120603. doi: https://doi.org/10.1103/PhysRevLett.105.120603.

Takens, F. (1981). Detecting strange attractors in turbulence. *Lecture Notes in Mathematics*, 898:366–381.

Talagrand, O. and Courtier, P. (1987). Variational assimilation of meteorological observations with the adjoint vorticity equation i: Theory. *Quarterly Journal of the Royal Meteorological Society*, 113:1311–1328.

Tang, D. Y., Dykstra, R., Hamilton, M., and Heckenberg, N. (1998). Observation of generalized synchronization of chaos in a driven chaotic system. *Physical Review E*, 57:5247–5251.

Teschl, G. (2010). *Ordinary Differential Equations and Dynamical Systems (Graduate Studies in Mathematics)*, vol. 140. American Mathematical Society. ISBN-13: 978-0821883280; ISBN-10: 0821883283.

Thomson, R. E. and Emery, W. J. (2014). *Data Analysis Methods in Physical Oceanography (Third Edition)*. Elsevier Science, Amsterdam.

Tikhonov, A. N. and Arsenin, V. Y. (1977). *Solutions of Ill-Posed Problems*. Wiley, New York.

Toth, B. (2011). *Computational Methods for Parameter Estimation in Nonlinear Models*. PhD thesis, University of California San Diego.

Toth, B. A., Kostuk, M., Meliza, C. D., Margoliash, D., and Abarbanel, H. D. I. (2011). Dynamical estimation of neuron and network properties i: variational methods. *Biological Cybernetics*, 105(3–4):217–237.

Trevisan, A., D'Isadoro, M., and Talagrand, O. (2010). Four dimensional variational assimilation in the unstable subspace. *Quarterly Journal of the Royal Meteorological Society*, 136:487–496.

Ty, A. J. A., Fang, Z., Gonzalez, R. A., Rozdeba, P. J., and Abarbanel, H. D. I. (2019). Machine learning of time series using time-delay embedding and precision annealing. *Neural Computation*, 31:2004–2024.

Vallis, G. K. (2017). *Atmosphere and Ocean Fluid Dynamics: Fundamentals and Large-Scale Circulation*. Cambridge University Press. doi: https://doi.org/10.101.1017/9781107588417.

Wachter, A. and Biegler, L. T. (2006). On the implementation of a primal-dual interior point filter line search algorithm for large-scale nonlinear programming. *Mathematical Programming*, 106(1):25–57.

Wendlandt, J. M. and Marsden, J. E. (1997). Mechanical integrators derived from a discrete variational principle. *Physica D: Nonlinear Phenomena*, 106:223–246.

Whartenby, W. (2012). *Methods for Data Assimilation in Chaotic Systems: Examples from Simple Geophysical Models*. PhD thesis, University of California San Diego.

Whartenby, W., Quinn, J., and Abarbanel, H. D. I. (2013). The number of required observations in data assimilation for a shallow water flow. *Monthly Weather Review*, 141:2502–2518.

Whittaker, E. T. and McCrae, S. W. (1988). *A Treatise on the Analytical Dynamics of Particles and Rigid Bodies*. Cambridge Mathematical Library. Cambridge University Press. First published, 1904; 4th ed., 1947.

Wisdom, J. and Holman, M. (1991). Symplectic maps for the n-body problem. *Astronomical Journal*, 102:1528–1538.

Ye, J. (2016). *Systematic Annealing Approach for Statistical Data Assimilation*. PhD thesis, University of California San Diego.

Ye, J., Kadakia, N., Rozdeba, P. J., Abarbanel, H. D. I., and Quinn, J. C. (2015a). Improved variational methods in statistical data assimilation. *Nonlinear Processes in Geophysics*, 22(2):205–213.

Ye, J., Rey, D., Kadakia, N., Eldridge, M., Morone, U. I., Rozdeba, P., and Abarbanel, H. (2015b). Systematic variational method for statistical nonlinear state and parameter estimation. *Physical Review E*, 92:052901.

Zhong, G. and Marsden, J. E. (1988). Lie-Poisson Hamilton-Jacobi theory and Lie-Poisson integrators. *Physics Letters A*, 133(3):134–139.

Zhu, C., Byrd, R. H., Lu, P., and Nocedal, J. (1997). Algorithm 778: L-bfgs-b: Fortran subroutines for large-scale bound-constrained optimization. *ACM Transactions on Mathematical Software (TOMS)*, 23(4):550–560.

Zinn-Justin, J. (2002). *Quantum Field Theory and Critical Phenomena*. Oxford University Press.

Index